Environmental Management in Proactive Commercial Firms

Lessons for Central Logistics Activities in the Department of Defense

Frank Camm

Prepared for the
Office of the Deputy Undersecretary of Defense for Environmental Security

National Defense Research Institute

RAND

The research described in this report was sponsored by the Office of the Deputy Undersecretary of Defense for Environmental Security. The research was conducted in RAND's National Defense Research Institute, a federally funded research and development center supported by the Office of the Secretary of Defense, the Joint Staff, the unified commands, and the defense agencies under Contract DASW01-01-C-0004.

Library of Congress Cataloging-in-Publication Data

Camm, Frank A., 1949-
  Environmental management in proactive commercial firms : lessons for central logistics activities in the Department of Defense / Frank Camm.
    p. cm.
    "MR1308."
    Includes bibliographical references.
    ISBN 0-8330-2958-4
    1. United States. Dept. of Defense—Procurement—Environmental aspects. 2. Defense contracts—United States—Environmental aspects. 3. Environmental management—United States. 4. United States. Dept. of Defense. I. Title.

UC263 .C3622 2001
355.6—dc21

2001019325

RAND is a nonprofit institution that helps improve policy and decisionmaking through research and analysis. RAND® is a registered trademark. RAND's publications do not necessarily reflect the opinions or policies of its research sponsors.

Published 2001 by RAND
1700 Main Street, P.O. Box 2138, Santa Monica, CA 90407-2138
1200 South Hayes Street, Arlington, VA 22202-5050
RAND URL: http://www.rand.org/
To order RAND documents or to obtain additional information, contact Distribution Services: Telephone: (310) 451-7002; Fax: (310) 451-6915; Internet: order@rand.org

In the past decade, a new approach to environmental management has emerged among innovative organizations and regulators, an approach that draws on broader efforts to induce organizational learning to improve production and management processes on a continuing basis. Innovative organizations are viewing environmental management not simply in terms of requirements to comply with specific rules and regulations in place today, but also in terms of ways to adjust product designs, production and delivery processes, and organizational behaviors over time to cut the total social costs associated with environmental emissions and, where possible, turn environmental issues to strategic advantage.

The Department of Defense (DoD) has participated in this new approach and wants to continue its role as a proactive innovator, seeking to improve the nation's environment in ways that are compatible with the DoD's primary responsibility to provide for the national defense. To maintain its awareness of analogous efforts among commercial firms, the Office of the Deputy Under Secretary of Defense for Environmental Security asked RAND to review new environmental management programs in the commercial sector relevant to weapon system design, provision of central logistics services, integration of defense activities on installations, and management of cleanup programs.

Early work on the project revealed that the DoD's general policies on environmental management were compatible with policies being developed by leading firms that have activities analogous to those in these four areas. In response, the Under Secretary asked RAND to focus on the effective *implementation* of these policies. To do this, RAND initiated case studies of environmental management in two leading firms relevant to each of the four policy areas. These case studies allowed the researchers to examine environmental policies in context and to understand how specific organizations have integrated these policies with their broader corporate cultures. It was found that such integration lies at the heart of successful implementation of environmental management policies. This report focuses on implementation in organizations whose activities are similar to those of central logistics services.

Related reports include the following:

- S. Resetar, F. Camm, and J. Drezner, *Environmental Management in Design: Lessons from Volvo and Hewlett-Packard for the Department of Defense,* MR-1009-OSD, RAND, 1998.

- J. Drezner and F. Camm, *Using Process Design to Improve DoD's Environmental Security Program: Remediation Program Management,* MR-1024-OSD, RAND, 1999.

This research was conducted for the Office of the Deputy Under Secretary of Defense for Environmental Security within the Acquisition and Technology Policy Center of RAND's National Defense Research Institute, a federally funded research and development center sponsored by the Office of the Secretary of Defense, the Joint Staff, the unified commands, and the defense agencies.

CONTENTS

FIGURE AND TABLES

Figure

Tables

SUMMARY

Central logistics activities in the Department of Defense (DoD) face environmental performance challenges very similar to those faced by industrial activities in the private sector. The regulatory environment is becoming more demanding as regulations grow in complexity, and in addition, persistent demands to reduce the size of the DoD "infrastructure" (however that is defined) have focused attention on any costs associated with that infrastructure, including the costs of meeting environmental goals.

Like many of its innovative commercial counterparts, the DoD has sought to take a proactive approach to environmental concerns. Rather than simply reacting to regulators' demands at the "end of the pipe," DoD policymakers have sought policies that are more positive and constructive. These policies try to motivate personnel to adjust environmentally relevant processes in ways that enhance the DoD's ability to pursue its core military mission while maintaining its standing as an environmentally responsible public agency that sets a good example for others in government and the private sector. But like many of their commercial counterparts, high-level decisionmakers in the DoD have found it difficult to implement proactive policies in ways that affect decisions made throughout the organization.

The Office of the Deputy Undersecretary of Defense for Environmental Security asked RAND to study how commercial firms recognized as having the best environmental management practices in the country have implemented those practices. Lessons from these firms should help the DoD improve its own implementation of a proactive approach to environmental management. This report addresses commercial implementation of environmental management practices that are relevant to the DoD's central logistics activities.

Activities often associated with proactive environmental management, in the DoD and elsewhere, include

- Designing general environmental management systems (EMSs), including metrics, to ensure integration.

- Training and motivating people.

- Providing tools and information to support the environmental mission.

- Promoting effective relationships with stakeholders

- Using ISO 14000, a series of proposed international guidelines that could become standards for environmental management practices, as a framework for implementing specific changes relevant to the issues above.[1]

In this study, RAND used corporate documents and a review of the secondary trade and academic literature to examine the methods that successful proactive firms have used to implement environmental management policies. The study yielded seven findings that are important to central logistics:

1. Lessons from environmental management in commercial settings can be applied to DoD central logistics activities. In wholesale-level transportation and supply, fuel consumption and packaging are particularly important environmental issues, but many other concerns also arise. Depot-level maintenance of defense systems raises the same variety of environmental issues that exist in the manufacture of comparable integrated electronics and aircraft systems, engines, and heavy equipment. These issues range from direct emissions to hazardous-waste generation. No one environmental concern dominates in central logistics. The sources of environmental problems and, by implication, the keys to solving them exist throughout the processes that produce central logistics services. They are, for practical purposes, an integral part of these processes.

---

[1]The first element of the series, ISO 14001, was approved in 1996.

Any effort to manage these concerns must recognize how inter-twined environmental issues are with the design and operation of logistics production processes.

2. A broad consensus is emerging from the many commercial-sector experiments in proactive environmental management that are under way today. Many commercial firms with production processes analogous to those in central logistics are experimenting with proactive environmental management. Each year, additional major firms find ways to look beyond simply complying with current regulations. They cite many reasons for doing this. As regulations have become more pervasive, more and more cost-effective opportunities to move away from end-of-the-pipe controls to pollution prevention present themselves. Planning before regulations are imposed gives an organization a better opportunity to develop a cost-effective response to such regulations. As global entities, large firms find it easier to develop a single, organizationwide policy than to worry, at the corporate level, about specific compliance issues in each locale. As highly visible parts of the communities where their facilities reside, they can affirm their status as responsible citizens by doing more than the regulations require. Within this consensus, each firm finds a way to tailor its environmental management approach to fit its own corporate culture. To the extent that the DoD wants to emulate these firms, it will need to fit its environmental programs to its own culture.

3. The environmental management policy that the DoD promotes today is broadly compatible with the consensus emerging from best commercial practice. The DoD's environmental policy highlights many of the points that proactive commercial firms emphasize: De-velop and maintain good working relationships with key stakehold-ers; develop clear organizational goals and metrics to support them; improve information systems that can be used to monitor perfor-mance; motivate the workforce and empower it with effective train-ing and analytic tools; support experimentation with innovative technologies; use affirmative acquisition policies to encourage sup-pliers to be more environmentally responsible; and so on. From this perspective, high-level environmental policymakers in the DoD al-ready support much of the consensus forming in the commercial sector.

4. The DoD should integrate environmental management with its core mission concerns. Integration makes connections in both directions. Moving corporate values into the realm of environmental management emphasizes the importance of going beyond compliance only to the extent that such a change is compatible with the corporate strategy. Moving an environmental perspective into the core production processes of the firm creates new opportunities for these processes to pursue the long-term corporate strategy. Effective integration is achieved when environmental concerns are no more and no less important than the other concerns in an organization. This view is compatible with the observation that environmental concerns are integrally intertwined with the core concerns of designing and operating central logistics production processes.

Such integration is natural in central logistics activities. The DoD is already using an integrated supply chain perspective to link logistics functions to one another and to the core military missions of the department. Agile combat support and velocity management are two innovative examples of this ongoing effort. Integrated supply chain management provides precisely the perspective needed to bring environmental concerns closer to the core concerns of logistics. It provides a natural way to ask, for example, where shortened cycle times are compatible with environmental and core military goals and where they are not. It provides a natural way to ask where pollution-prevention activities can reduce emissions and waste in ways that reduce systemwide operating costs inside the DoD. In sum, current DoD thinking about integrated logistics services provides a natural conduit for promoting the integration of environmental and core concerns.

5. A formal environmental management program can increase the likelihood that implementation of proactive environmental management will succeed. Even as a proactive organization integrates environmental management and management of its core mission, it relies on a formal environmental program to maintain focus. The program cannot directly do the things that make the program succeed; core production activities must do that. Rather, the program maintains visibility over environmental issues to ensure that core production activities take them seriously.

A "formal program" typically includes managers whose sole responsibility is environmental policy at the corporate, business-unit, and plant level. It includes a staff to support these managers and a career track to develop staff and managers over time. This career track provides the specialists who serve on cross-functional teams.

A formal program also typically includes a monitoring system that develops goals and objectives at each level in the organization, develops metrics to monitor performance against those objectives, and routinely reports information on those metrics to the leaders responsible for core production activities at each level of the organization. Environmental specialists help the core production personnel develop these goals, objectives, and metrics, and the incentives to support them, but the core personnel are ultimately responsible for them. The monitoring system provides a way for the organization's leaders to see how well the core activities execute against their own plans and to reward them accordingly. The environmental program provides support and advice in this effort, but the environmental specialists do not make decisions about production activities, investments, or rewards.

Among other things, a formal program monitors the development and execution of proactive environmental initiatives. It monitors progress against the milestones developed for each initiative, and it also helps structure the set of initiatives itself to promote learning over time. Successful firms typically try easy, lower-risk initiatives first, i.e., actions farther from core production activities. These firms build on success: They use early success to maintain corporate support for more work and to develop lessons that the organization can use to attack more difficult challenges. These firms "seed" initiatives at the beginning to promote such learning in each of their business units and key production areas.

6. Total Quality Management (TQM) provides useful formal templates that the DoD can use to verify its approach to implementing proactive environmental management. Many proactive firms emphasize the inherent compatibility of TQM and proactive environmental management, particularly pollution prevention. TQM and pollution prevention share the goal of eliminating waste, i.e., any activity that does not contribute value for the final customer or stakeholder. Such a bottom-up perspective is especially appropriate in central logistics activities, where waste exists throughout the supply chain and can be flushed out only by a systematic effort to review each process and its contribution to final customer value. In this setting, pollution prevention becomes just one point of focus for the TQM continuous improvement process at all levels. Seen in this way, TQM creates another opportunity to link an organization's environmental management processes with its core mission processes.

Several formal TQM-based templates provide the basis for monitoring an organization's efforts to create and maintain a formal environmental management program like that described above, from the top down. ISO 9000 and ISO 14000 offer compatible frameworks for creating a quality-based organization and making sure that it reflects environmental management concerns effectively. Used together, these standards promote the kind of integration discussed above. The Baldrige Award criteria and Total Quality Environmental Management matrix developed by the Council of Great Lakes Industries provide mutually compatible frameworks for moving beyond the requirements of the ISO templates. Again, used together, they promote integration in an organization seeking best-of-class status, not just in environmental management, but across the board.

7. Effective, proactive environmental management in central logistics activities could help lead the DoD toward broad acceptance of TQM. As noted above, a common theme in commercial experience is the need to develop an approach to environmental management that is compatible with the core values of the organization. Firms that have adopted TQM naturally saw it as a tool they could use to pursue proactive environmental management. In fact, some saw the application of TQM to environmental issues precisely as a natural way to justify environmental efforts in their corporate structures and hence a natural way to integrate environmental concerns with the broader concerns of the firm. TQM became the natural vehicle to use to promote integration.

The DoD does not maintain a quality-based infrastructure. The functional organizations in the services that maintain TQM skills can point to many successes, but they have not penetrated the core activities of the services. As the DoD reviews its infrastructure to squeeze out the dollars it needs to recapitalize the force in the new century, it will have to achieve dramatic improvements in performance. TQM and related reengineering perspectives are the only concepts that have yielded such improvements in commercial firms. If the DoD is to improve its infrastructure, as it believes it must, TQM will have to move toward center stage soon.

DoD environmental policymakers can take advantage of that move by linking environmental management to TQM in a way that puts them in common cause with innovators elsewhere in the department. TQM offers the potential for creating the centerpiece for a coalition of innovators who work together to change the DoD's approach to managing the infrastructure. Existing proactive environmental policy in the DoD has already created an opening. The environment might actually serve as an ideal place to open wedges to work toward such an infrastructure. Because environmental concerns are more distant from the core concerns of the DoD than most infrastructure issues, they offer a low-risk place to test TQM concepts that could be applied elsewhere in the department if they prove to be successful.

ACKNOWLEDGMENTS

Gary D. Vest, Principal Assistant Deputy Under Secretary of Defense (Environmental Security), suggested and supported RAND's study of proactive environmental management practices in commercial firms, of which this report is one product. Patrick Meehan, Jr., Director of Program Integration for the Deputy Under Secretary of Defense (Environmental Security), was the program officer for this broader study. Both provided useful suggestions and feedback, as well as ready access to relevant staffs in the Office of the Secretary of Defense and the military departments.

Within RAND, I worked closely with colleagues Jeffrey Drezner, Beth Lachman, and Susan Resetar on RAND's broader study of proactive environmental practices. We designed that study together and helped each other execute specific parts of it. Jeff, Beth, and Susan were all very helpful to me in my efforts to gather and synthesize materials relevant to central logistics. D. J. Peterson's reviews of all of the products of the joint effort, including this one, have significantly improved their quality. Quality and defense logistics specialists at RAND helped me link key ideas about proactive environmental management to lean production and, more generally, integrated supply chain management. I thank especially Irv Cohen, Shan Cretin, Jeff Luck, Nancy Moore, Ray Pyles, and Hy Shulman. Janet DeLand ably edited the document, and Christopher Kelly oversaw final editorial production.

I thank them all but retain full responsibility for the accuracy and analytic soundness of the material presented here.

ACRONYMS

ABC   Activity-based costing

BRAC   Base Realignment and Closure

CEO   Chief executive officer

CERCLA   Comprehensive Environmental Response, Compensation, and Liability Act of 1980 (Superfund)

CFC   Chlorofluorocarbon

DoD   Department of Defense

EHS   Environmental, health, and safety

EMAS   Eco-Management and Audit Scheme

EMP   Environmental Master Plan (at IBM)

EMS   Environmental management system

EPA   Environmental Protection Agency

GEMI   Global Environmental Management Initiative

HCFC   Hydrochlorofluorocarbon

ISO   International Standards Organization

NOV   Notice of violation

POL   Petroleum, oil, and lubricants

PRP   Primary responsible party

RCRA   Resource Conservation and Recovery Act of 1976

SPO   System Program Office

TQEM   Total Quality Environmental Management

TQM   Total Quality Management

TRI   Toxic Release Inventory

Chapter One

INTRODUCTION

The Department of Defense (DoD) refers to its central logistics activities as industrial activities for good reason. The inside of a maintenance depot, materiel warehouse, or major transit point looks very much like its industrial counterpart in the private sector. Overhauls in maintenance facilities consist of breaking down and then remanufacturing large, sophisticated structures that use complex electronic parts. DoD warehouses store items very much like those in commercial warehouses, and they use similar technologies to manage them. And the facilities that load and support trucks, transport aircraft, and ships execute the same tasks their commercial counterparts do.

It is therefore not surprising that DoD central logistics activities face many of the same environmental challenges faced by industrial activities in the commercial sector. Over the past decade, environmental concerns have become more visible in the DoD for many of the same reasons they have received attention in commercial circles during the same time period. The DoD and private firms now face essentially a very similar regulatory environment. That environment is becoming more demanding as regulations grow in complexity. In addition, persistent demands to reduce the size of the DoD "infrastructure" (however that is defined) have focused attention on any infrastructure-related costs, including those of meeting environmental goals.

Like many of its innovative commercial counterparts, the DoD has sought to take a proactive approach to its environmental concerns. Rather than simply reacting to regulators' demands at the "end of the pipe," the DoD has sought more positive and constructive policies.

These policies try to motivate personnel to adjust environmentally relevant processes in ways that enhance the DoD's ability to pursue its core military mission while maintaining its standing as an environmentally responsible public agency that sets a good example for others in both the government and the private sector. But like many of their commercial counterparts, high-level decisionmakers in the DoD have found it difficult to implement proactive policies in ways that affect decisions throughout the organization.

The Office of the Deputy Under Secretary of Defense for Environmental Security asked RAND to study the environmental management practices of commercial firms that are recognized as having the best practices of this kind in the country. The findings should provide lessons that the DoD can use to improve its own environmental management practices.

ANALYTIC APPROACH

RAND first reviewed the DoD's environmental management program. On the basis of material developed through reading high-level DoD documents, interviewing environmental officials in the armed services, and visiting several representative bases, RAND and the DoD agreed about two points: First, the DoD's high-level environmental management *policy* already reflects many of the practices that are proving successful in commercial firms. Thus, the DoD's primary challenge is less one of knowing what to do than one of understanding how to *implement* environmental policy in a very large organization. As a result, the DoD needed to be most concerned about how proactive commercial firms had implemented the key elements of their environmental management programs.

Second, among the many environmental activities the DoD pursues, four are particularly important:

- Designing new major weapon systems so that they will have cost-effective environmental performance levels over their lifetimes.

- Managing the industrial processes in central logistics activities so that they balance military, environmental, and cost constraints appropriately in the support of existing weapon systems.

- Managing the many diverse, environmentally relevant activities on installations in an integrated way to comply with current regulations and prevent future pollution cost-effectively.

- Managing the programs responsible for remediating waste disposal sites on active bases, bases being closed through Base Realignment and Closure (BRAC) reviews, and other sites where the DoD is a primary responsible party (PRP).

RAND focused on these four areas. Related findings are documented in two companion reports.[1]

This report focuses on the second area, management of environmental processes in central logistics activities. It uses an in-depth review of the recent academic, policy, and trade literature to construct general descriptions of best practices in the commercial sector relevant to central logistics.

The best commercial firms are developing and implementing very innovative practices, and like the DoD, they are finding that implementation is more challenging than policy development. Activities normally associated with implementing proactive environmental management include

- Designing general environmental management systems, including metrics, to ensure integration.

- Training and motivating people.

- Providing tools and information to support the environmental mission.

- Promoting effective relationships with relevant stakeholders.

- Using ISO 14000 as a framework for implementing specific changes relevant to the issues above.[2]

---

[1]Resetar, Camm, and Drezner, 1998; and Drezner and Camm, 1999.

[2]ISO 14000 is a series of proposed international guidelines that could become standards for environmental management practices and could shape the DoD's regulatory environment in the future. The first element of the series, ISO 14001, was approved in 1996. Chapter Three discusses ISO 14000 at greater length.

A NOTE ON TOTAL QUALITY MANAGEMENT

The discussion that follows makes numerous references to Total Quality Management (TQM), a family of management techniques initially developed in the United States in the 1940s, then refined and expanded in Japan during the 1950s, 1960s, and 1970s, and finally rediscovered in the United States in the 1980s. The many American and Japanese consultants promoting the use of TQM have worked hard to differentiate their products and, in the process, have generated a great deal of confusion about what exactly TQM is.[3] At its heart, TQM emphasizes three elements:[4]

- Identify the organization's customers and what they want, now and in the future.

- Identify the processes in the organization that serve the customers and eliminate as much waste—activity that does not add value for the customers—as possible from these processes.

- Monitor performance against the first two goals and continually improve that performance over time.

"Quality" refers to anything the customer values, and TQM seeks to eliminate any activity that does not contribute to that quality. References in the text point to management techniques in proactive organizations that embody the three elements above.

ORGANIZATION OF THIS REPORT

Chapter Two provides the basic context for the study by (1) explaining the link between central logistics activities in the DoD and particular industrial processes in commercial firms, and (2) explaining important parallels between proactive environmental policy and the "integrated supply chain," another concept relevant to innovative logistics practice today. Chapter Two defines central logistics to include all depot-level transportation, supply, maintenance, and logistics control activities. Maintenance includes overhaul, or re-

---

[3]A well-written introduction to TQM that distinguishes among its variants without getting lost in them is Dobyns and Crawford-Mason, 1991.

[4]Levine and Luck, 1994; see also Womack and Jones, 1996.

manufacturing, activities, as well as repair activities. For our purposes, remanufacturing looks a great deal like manufacturing. Successful commercial firms are applying the concept of an integrated supply chain to link their logistics activities together and to their core business activities. Chapter Two reviews this concept and its relevance to environmental management. It then reviews the environmental concerns that arise in commercial transportation, supply, and manufacturing activities, giving special attention to firms that manufacture aerospace systems, heavy machinery, and sophisticated integrated electronics systems.

Chapter Three synthesizes information, primarily from the 1990s, on commercial environmental management practices in transportation, supply, and manufacturing activities. This synthesis shows that a strong consensus is forming around a set of proactive environmental practices whose ultimate goal is to integrate environmental management with the management of the core activities in an organization. The implementation approach that proactive firms rely on most heavily is that associated with TQM, which now offers formal quality templates that significantly increase the likelihood that a firm will be successful in its efforts to implement proactive environmental management. Chapter Three presents an overview of relevant formal quality templates.

Chapter Four focuses on the ISO 14001 quality template, reviewing the benefits that it offers commercial firms and comparing them with those the DoD might obtain. It then reviews issues that arise in the implementation of formal efforts to use ISO 14001 in the commercial sector. Ford's and IBM's recent successful global registrations to ISO 14001 are discussed to illustrate these benefits and implementation challenges in two different settings.

Chapter Five presents conclusions and recommendations relevant to environmental management in the DoD. The analysis concludes that the DoD already supports a proactive environmental management *policy* very much like that being tested in the commercial sector and that the DoD could follow the commercial lead toward *implementing* this policy. To do this, the DoD could benefit greatly from adopting formal TQM-based tools. Moreover, the use of these tools in the environmental management of central logistics activities could help promote their broader application throughout the DoD.

Chapter Two

ENVIRONMENTAL ISSUES IN CENTRAL LOGISTICS ACTIVITIES

Central logistics consists of depot-level, or wholesale, transportation, supply, and maintenance activities conducted in the DoD. Although the DoD has traditionally treated these activities as separable functions, somewhat removed from the core combat missions of the department, innovative commercial firms view these same activities very differently.[1] Using an integrated supply chain perspective, they link these functions to one another and to the core business concerns of the firm. The logic used to integrate logistics functions with core business concerns is very similar to that used by many innovative firms to integrate environmental concerns with core business concerns. DoD logisticians are learning how to use this integrated supply chain perspective as they implement variations on it, such as agile combat support and velocity management.[2] They could easily

---

[1]We generally use language that is compatible with the current quality-based approach to organizational management. A *function* is an area of expertise within an organization, e.g., logistics or environmental control. A *process* is a production activity that transforms a clear set of inputs into a clear set of outputs, e.g., the series of actions required to fulfill an order. Although many organizations are organized around their functions—areas of expertise through which employees rise during their careers—processes typically cut across functions, because many different skills are important to the production activities of an organization. These distinctions are important to the primary theme of this report, the need to integrate across *functions* to ensure that production *processes* handle environmental issues properly.

[2]The Air Force agile combat support program, originally called *lean logistics,* seeks to apply the basic principles of lean production to its logistics support system. The Army velocity management program focuses on reducing "order and ship time," the time it takes to move an order from the field to the depot and for a part to move from the

extend such thinking to reflect the environmental concerns of the department. Therefore, it is important to clarify the environmental challenges facing logistics operations.

This chapter briefly reviews the basic elements of central logistics in the DoD. It then explains the integrated supply chain perspective and its relevance to traditional logistics functions and environmental issues. Finally, it reviews the traditional logistics functions in greater detail and identifies the most important environmental concerns associated with each, drawing on the extensive secondary literature on relevant experience in proactive commercial firms.

CENTRAL LOGISTICS IN THE DoD

Central logistics in the DoD handles four kinds of activities: transportation, supply, maintenance, and control. *Transportation* includes the organic and contractor movement of personnel and materiel by truck, rail, air, or water. It also typically includes concerns about packaging associated with transporting materiel. Materiel can consist of anything from a major weapon system to subcomponents and consumable parts, commercial and industrial-type equipment, bulk chemicals and fuels, food and medical supplies, and office and facility maintenance supplies. *Supply* refers to the procurement of such materiel and services, typically from private-sector sources; warehousing of materiel in organic or contractor facilities; and distribution of materiel from suppliers to central warehouses and from central warehouses to defense activities, where it may be used or stored for future, local use. *Maintenance* is the organic and contractor overhaul, or remanufacturing, and repair of major systems and subcomponents and the maintenance of equipment, facilities, and other physical assets required to provide such overhaul and repair. Maintenance typically includes on-site management of materiel required for immediate support of overhaul and repair, including in-process inventories. *Control* consists of the information systems used to track the status of all organic and contractor logistics activities, accounting systems and metrics to support such tracking, and planning and programming functions to support investment deci-

---

depot to the point in the field where the order was initiated. See, for example, Dumond et al., 1994; Ramey, 1999.

sions and other resource management activities relevant to the management of central logistics activities.

These activities occur in the individual armed services and in the Defense Logistics Agency and other defense agencies. The services and joint commands act as the consumers. The activities are typically provided with organic assets, but the services and agencies are under growing pressure to seek contractor sources. Central logistics has, in fact, been a target of recent efforts to outsource defense activities.[3] Some of the recent reforms aimed at moving the DoD toward greater use of commercial practices envision using such changes to integrate central logistics more securely with the best capabilities available in the private-sector industrial base. Hence, central logistics is a logical place to think about implementation of commercial-style environmental management practices.

INTEGRATED SUPPLY CHAIN MANAGEMENT

Total Quality Management (TQM) and reengineering efforts center on identifying relevant *customers* and then identifying the *processes* that an organization uses to serve those customers.[4] Logistics processes ensure that an organization receives the inputs required to produce its primary products, gets those products to its customers, and continues to support those products cost-effectively over their lifetimes.

Thus, logisticians have created the concept of an integrated supply chain that links all the relevant processes, from the production of the raw materials that an organization uses to create a product to the ways its customers use the product it provides.[5] They often speak of "walking the box," by which they mean tracing a physical part, or "box," from its point of origin to its final destination and analyzing each process, "end to end," that affects the box on its journey. TQM

---

[3]Commission on Roles and Missions of the Armed Forces, 1995; Defense Science Board, 1996a, 1996b.

[4]For an overview of the many approaches and how they relate to one another, see Levine and Luck, 1994.

[5]For an excellent discussion, with illustrations, see Womack and Jones, 1996, Chap. 2. Also see Rose and Sharman, 1989, pp. 29–43.

and reengineering seek to identify innovations in these processes that strip the waste—anything that does not add value for the customer—out of every element of this supply chain.[6] Judging every process in terms of its contribution to the final customer integrates the supply chain and invites life-cycle assessment of the processes from the customer's point of view.

In the DoD, the central logistics activity that has received the most such attention is the maintenance of secondary items sent to a depot for repair and then returned to service.[7] The relevant customer for this activity is the organization that uses secondary items to keep its major weapon systems operational. This is how this activity looks end to end: (1) An item on a weapon system is first identified as faulty. It is typically screened by the customer's own maintenance facility to determine whether it can be repaired locally. (2) If the local facility cannot repair it, the customer generates a demand on the central logistics system: It ships the item to a depot for repair. (3) The depot repairs it but may have to wait for parts needed for the repair; these may have to be procured or repaired in their own right. (4) When the repair is completed, the depot either ships the item to a supply depot for storage until it is needed or returns it to service immediately.

In sum, the item starts at a military facility, is transported to a central repair depot, is maintained using supplies as needed, and is transported to its final destination. Transportation, supply, and maintenance all play direct roles; control provides the information systems that control this flow and conducts the planning and programming required to allocate resources to these functions as needed to sustain or improve the flow. An end-to-end mapping of this very simple repair activity shows how all the key elements of central logistics come to bear on it.

---

[6]"Waste" in a quality context is different from material waste as normally understood in an environmental setting. But quality advocates often draw a parallel between any environmental effluent and waste from a quality perspective. Effluents do not add value to the final customer. Hence they are all waste in a quality context. That is why so many view TQM as a suitable vehicle for identifying and reducing environmental effluents. This report uses the term *waste* in both senses, relying on clarifications or the context to provide the intended meaning.

[7]As noted earlier, the Army does this in its velocity management program, and the Air Force in its agile combat support program.

From a customer's point of view, three things about this activity matter:

- The customer needs an item to replace the faulty item as soon as possible to keep its major systems operational.

- The customer wants the replacement item to work as predicted when it is installed in a major system.

- The customer wants to pay as little as possible to obtain the desired result.

In the past, the DoD pursued these goals by maintaining massive inventories that enabled a quick replacement whenever needed and additional replacements if the first ones proved faulty. However, the DoD is now taking advantage of innovations in the commercial sector based on integrated supply chain management that allow it to substitute information and transportation, which are falling in cost, for increasingly expensive parts.[8]

These efforts have yielded the following insights:

- Speeding up the end-to-end process, i.e., "reducing the cycle time," allows the DoD to reduce the inventories required to fill "pipelines" of reparable items in transit or repair. Smaller inventories mean lower costs.

- Speeding up the end-to-end process reveals problems in the total process more quickly. The ability to resolve problems without long delays reduces the inventories that must be held in anticipation of such problems.

- Spending money on faster transportation, better management of the "retrograde pipeline" that returns items from the field for repair, more maintenance capacity, or more subpart inventory in the maintenance shop can cut the need for total inventories of parts enough to more than justify the added expense.

- Prioritization of repair and distribution to provide service first to the customers in greatest need can reduce the need for inventories and repair capacity. Savings on inventories and repair

---

[8]For a discussion of this point, see Girardini et al., 1995.

capacity can pay for information systems that improve prioritization.

These observations all offer examples of another, more basic insight: When an organization views an integrated supply chain from the perspective of the customer, it can make many cost-effective changes in individual parts of the chain that may not look cost-effective if viewed only in the context of an individual part.[9] Viewing the chain as a whole and understanding how policy changes affect all parts of it provide unique opportunities for improvement.

These insights come from "walking the box" through the processes that support secondary item repair. Analogous insights could come from focusing on a chemical solvent rather than a reparable part. In this case, the DoD buys a chemical, stores it until needed, applies it in a maintenance process, loses a portion of it through emissions to the air and water, spills a portion of it in an unanticipated accident, recovers a portion of it through recycling for reuse, and ultimately treats or disposes of the portion that is no longer usable. RAND analysis has shown that factors affecting one part of this life cycle can easily influence policy decisions in other parts of the life cycle.[10] Using the integrated view here yields the following insights:

- Changing a maintenance process to use less of a chemical, recycling the chemical, or treating it before disposal to reduce its volume or toxicity can cut disposal costs. The value of these changes rises with increases in the cost of disposal, associated with the actual list price of disposal services, liability associated with property damage or personal injury after disposal, or requirements for storage and remediation under RCRA or CERCLA.[11]

- Stricter sanctions and bad publicity associated with air and water emissions and accidental spills increase incentives to limit the quantity of the chemical used or tighten control over its use.

---

[9]Spriggs, 1994, p. 216, provides useful illustrations of this insight.

[10]For a more detailed discussion of these points, see Wolf and Camm, 1987.

[11]The Resource Conservation and Recovery Act of 1976, and the Comprehensive Environmental Response, Compensation, and Liability Act of 1980 (Superfund).

- Reducing cycle times for transportation of the chemical and its use in repair can dramatically reduce the quantities of the chemical needed in the inventory, thereby responding effectively to incentives created by environmental concerns.

- Higher permit costs, along with increases in administrative paperwork, operators' time, and interaction with regulators and other external groups, as well as the need for senior management to focus on concerns about regulation, liability, and bad publicity, can potentially increase the cost of using the chemical enough to warrant devising a maintenance approach that does not require the chemical at all, even if the direct costs of this alternative look high.

- Uncertainty about the regulations and liability rules that will be imposed on a chemical over its lifetime can likewise potentially impose large enough costs to warrant an approach to maintenance that does not require the chemical at all, even if the direct costs of this alternative look high.

Again, these insights confirm the basic insight that viewing the chain as a whole and understanding how policy changes affect all parts of it provide opportunities for improvement that can be obtained only with an integrated view. This insight restates the fundamental lesson that to be effective, environmental management must integrate environmental concerns with the core business concerns of an organization.[12] It must identify the total costs and benefits associated with environmental decisions throughout the organization. It must then use such information to support cost-benefit analysis of these decisions that is compatible with the cost-benefit analysis the organization uses to address its core business concerns. In this sense, an integrated approach to environmental management is

---

[12]This idea appears again and again in the literature and in our interviews. Fred Ellerbusch, Director of Corporate Health, Safety, and Environmental Affairs at Rhône-Poulenc, put it very simply: "Optimally, environmental concerns 'shouldn't be any more or any less important than any other. In full integration, [the environment] is not an afterthought or a forethought'" (quoted in Kirschner, 1996, pp. 19–20). Chapter Three discusses this idea in greater depth.

completely compatible with integrated supply chain management. The two perfectly complement and support one another.[13]

ENVIRONMENTAL ISSUES IN CENTRAL LOGISTICS

Integrated logistics can potentially have large payoffs in reducing the environmental problems associated with logistics. These problems are still most often stated in functional terms. The basic environmental concerns that arise in each traditional logistics function are reviewed below. For simplicity's sake, organic provision of these activities is addressed first, and the discussion then turns to questions that arise only in a contract setting. Table 2.1 summarizes the discussion. Even though the problems are stated one function at a time, the solutions often come from viewing any particular function in a broader system setting.[14]

Transportation

Most discussions of environmental concerns with regard to transportation as a logistics function focus on fuel consumption, particularly in trucking. Fuel consumption is important for two reasons: (1) The burning of fuel generates traditional air pollutants, especially sulfur dioxide, nitrogen oxides, carbon monoxide, particulate matter, and lead. (2) Fuel burning consumes a nonrenewable natural resource. These discussions implicitly view natural resource use as a relevant environmental concern, a perspective that is more important today in Europe than in the United States. The Europeans also increasingly add carbon dioxide as a serious threat to the global climate. All hydrocarbon fuels generate carbon dioxide as a product of combustion.

Most of the solutions offered for these problems point to improving fuel efficiency. Options include switching from truck to rail and from

---

[13]For a concrete example of how a commercial firm uses integrated supply chain management to design an environmental management program, see Rank Xerox, 1995.

[14]This discussion draws heavily on Cairncross, 1992; European Recovery and Recycling Association, 1991; Grogan, 1996; Paveley, 1992; Penman and Stock, 1994; Stillwell et al., 1991; Stock, 1992; Turbide, 1995; and Ummenhofer, 1995.

Table 2.1

Central Logistics Functions and Associated Environmental Concerns

Logistics Function	Environmental Concerns
Transportation	Energy consumption Air emissions Noise Waste disposal Spills and accidents Runoff from roads Congestion and road wear
Supply	Energy consumption Air emissions Hazardous-waste disposal Nonhazardous-waste disposal Normal spills and emissions Serious spills, fires
Maintenance	Direct emissions and small spills Hazardous-waste disposal Nonhazardous-waste disposal Serious spills, fires Energy consumption
Control	Compliance Programming and budgeting Information for integration and improvement
Contracting	Environmental effects of producing the products bought How products bought affect the buyer's environmental performance

internal combustion to electricity, acquiring vehicles with more efficient engines or more aerodynamic shapes, enforcing preventive maintenance more completely, using fuel additives to improve fuel efficiency, and training personnel to operate vehicles to improve fuel efficiency. They also suggest locating facilities to reduce the need for fuel consumption, tightening links in the logistics system to reduce the amount of materiel that must be shipped, and optimizing trip loads to reduce fuel consumption per pound of freight.

These last two options conflict with one another. Tightening of links typically means cutting cycle times by shipping items just in time, i.e., as needed; trip load optimization typically means accumulating full loads before a trip is generated. Some option between the two is likely to perform best, from an environmental perspective.

After fuel consumption, transporters worry most about noise, particularly from aircraft engines and from the road rumble, air brakes, and engines of trucks. Solutions typically call for quieter engines and vehicles or changing flight paths to reduce the disturbance to sensitive areas.

In transportation, waste disposal arises mainly with regard to packaging and pallets, although used tires and petroleum, oil, and lubricants (POL) also pose a challenge. The use of nonreusable packaging and pallets consumes valuable resources (although often renewable resources) and creates waste that must be managed.

The preferred solution is to reduce the amount of packaging and pallets needed. Once that is done, proactive firms seek to use reusable packaging and pallets. A formal reverse logistics program is then needed to return these items to their origin, typically on the backhaul segment of a trucking run. Agile combat support is compatible with such an approach because it returns a reparable part to the originating base in the same box that was used to ship it to the repair site; in its most complete form, agile combat support ships the *same* item (now repaired) to the originating base. The automobile industry has used reusable packaging and pallets to protect shipments for many years and has even conformed these to the trucks that carry them.

Many firms also use packaging that can be recycled, rather than composite packaging materials that combine paper and plastic elements. The basic principle that recycling works only with segregated waste requires a return to simpler materials. It also requires maintenance of a recycling infrastructure, which many firms choose to support by using packaging and pallets made from recycled materials. Alternatively, firms use packaging and pallets that can be ground up and composted, resulting in a product that can be used for grounds maintenance or burned cleanly, leaving only biodegradable ash for final disposal. Again, this requires the use of packaging made only of organic, biodegradable material.

Trucks and aircraft also generate waste when they are washed down during routine operational maintenance. Firms should use biodegradable cleaners and, ideally, should recycle, remove and capture effluent, and reuse wash water.

Waste disposal on ships raises different concerns. Plastic waste can now be found everywhere on the surface of the oceans, and chemicals can easily spill overboard. Solutions focus on tight control over shipboard operations to limit disposal at sea to organic waste and to hold all other waste for return to shore.

Spills and accidental releases are a serious concern in the transportation of hazardous materials. Responsible Care, the environmental code of the Chemical Manufacturers Association and its members, maintains a careful set of standards that all chemical firms have agreed to adopt in their transportation practices. Many firms go beyond these standards. One way to do this is to retain responsibility for the design of vehicles used to convey hazardous materials and even to retain in house the capability to transport such material. Another solution is to locate activities so that there will be less need to ship hazardous cargoes.

Natural resource damage also occurs in transportation but does not receive a lot of attention. POL washes off from paved surfaces and endangers sensitive tidal waters, e.g., those of Chesapeake Bay and San Francisco Bay, but proper vehicle maintenance can somewhat limit this runoff. Operation of ships can disrupt sensitive animal populations; the obvious solution is to avoid areas where these populations are at risk.

Finally, heavy use of roads contributes to congestion and road wear. Trucks account for much less congestion than cars do, even in Europe, where trucks are used more heavily for industrial shipments than in the United States But trucks account for a larger share of road wear than cars do, although of course, trucks pay fees that reflect this higher cost. Nonetheless, reduced shipments could reduce both congestion and road wear.

These environmental concerns raise several points that are relevant to all the logistics functions discussed below:

- Few environmental concerns evoke all-or-nothing solutions. Rather, the solutions call for substitution of rail for trucks, reductions in the amount of packaging used, or use of recyclable materials, where appropriate. Implicitly, such solutions assume a standard of appropriateness, even though they do not state the standard. To transform such solutions into clear guidance requires a clear definition of the standards underlying them.

- These solutions can easily affect the performance of logistics services relevant to the final customer. Changing the location of depots or routes of flight can easily affect the logistics system's ability to serve the customer. Thus, the standards of appropriateness must clearly address how important it is to be able to serve the final customer relative to other goals. It often helps to engage the customer in decisions that affect performance, and an integrated supply chain perspective helps link all effects to the customer.

- The regulatory environment stands behind these solutions but need not dictate them. For example, regulatory differences between the United States and Europe could point to different logistics practices in the two regions, a common logistics policy to simplify oversight within a multinational organization, or an approach to logistics that uses information on current regulation in each area to predict what the organization will face in the future in all areas. Differences between jurisdictions in the United States raise similar issues.

- Some solutions address effects that are not covered by regulation and that do not really directly affect the organization in question. Warehouses rarely get credit or blame for the ways their practices affect emissions from central powerplants. Adjusting maintenance policies to reduce runoff into Chesapeake Bay will normally not affect an organization's regulatory performance, although it can affect the organization's relationship with the regulator. Changing shipment policies to reduce road congestion or road wear will generate benefits that accrue primarily to parties that the organization cannot even name, much less negotiate with. These solutions reflect the full reach of the integrated supply chain and raise questions about where an organization should impose boundaries on that chain to define how integrated its own environmental policy should be.

Supply

Warehousing raises the most important environmental issues relevant to the supply function. Within warehousing, fuel consumption once again gets the most attention. The issues here, however, are not nearly as important as they are for the transportation function; transportation consumes much more fuel than warehousing and will do so for the foreseeable future.

Within the warehouse, heating, cooling, and lighting consume energy. These activities raise traditional concerns about fuel choice, efficiency of the fuel source, proper insulation, and exploitation of waste heat and district heating where possible. Movement of materials within the warehouse also consumes energy. Proper layout and planning can reduce the number of operations required to get materials in and out. And use of battery-operated electric vehicles allows reliance on efficient central powerplants rather than small internal combustion engines.

Reduced fuel use decreases emissions of the traditional air pollutants discussed above, and also of carbon dioxide. Other air pollutants associated with warehouses are chlorofluorocarbons (CFCs) used in chillers and, to a much lesser extent, halons used as fire extinguishants. Actions taken in response to the Montreal Protocol have largely displaced these chemicals as sources of concern. To the extent that hyrdochlorofluorocarbons (HCFCs) have replaced CFCs in chillers, another round of substitution can be expected as stratospheric ozone protection tightens. The ultimate solution will be to find alternative forms of chillers that do not rely on ozone depleters of any kind.

Packaging and pallets are the primary sources of nonhazardous waste in warehouses. The concerns and solutions discussed above apply here as well.

Warehouse operation per se generates very little hazardous waste. But warehouses store substantial quantities of hazardous materials, usually in transit from a supplier to a user, and some warehouses hold hazardous wastes for ultimate recycling, treatment, or disposal. Such hazardous-material and waste problems are effectively generated elsewhere and must be resolved at their sources.

Warehouses must maintain an effective management program to ensure proper labeling and rotation of stock and compliance with regulations on inventory management and storage times before treatment or disposal. Minimizing loss through obsolescence or other sources of shrinkage reduces the total requirement for chemicals in the organization.

Warehouses holding hazardous materials or waste should also apply standard industry practices to limit unwanted emissions, spills, or accidents and to limit the effects of such incidents when they occur. Some spillage and emission are inevitable—storage tanks leak, and emissions can occur when a hazardous material is transferred from one container to another or from a tank to a vehicle or container for shipment. Solutions include the replacement of corroded tanks and the placement of tanks so that they can easily be inspected for leaks; use of nitrogen to displace water vapor from half-empty tanks; and barriers and catchment equipment to capture emissions when they do occur.[15]

Large warehouses with substantial inventories of hazardous materials always face the danger of an uncontained spill or fire. Such events inevitably require outside assistance to resolve. Advance coordination with the relevant firefighting and hazardous-material control teams is now standard industry practice. This coordination includes sharing information on the chemicals present, although concerns about trade secrets still limit such sharing. Larger facilities also develop formal evacuation and emergency response plans, typically in coordination with local community groups and authorities.

Maintenance

Maintenance includes a broad range of industrial activities very similar to those commonly found in manufacturing. In fact, system overhaul—or remanufacturing, as it is more commonly called in the commercial sector—often occurs on the same assembly lines where manufacturing occurs.

---

[15]Table 3.2 in Chapter Three illustrates the complexity of the issues that must be considered when seeking solutions to even something as simple as a leaking storage tank.

Remanufacturing adds some steps to a standard manufacturing process. It starts with the disassembling of a system and the inspection of its constituent parts to determine which ones need to be replaced and which ones can be refurbished for continued use. Then some parts are directed to other processes for refurbishment, new parts are ordered to replace those that have been condemned, and a manufacturing program is scheduled that will reassemble the original system, using as many of the original parts as possible, or dispersing the original parts to supply bins and reassembling a new system from scratch. Once the new system is scheduled, manufacturing begins as it normally would if the original system had not been returned.

Repair involves a much more limited set of activities, but in the end, a continuum exists between simple repair and complete remanufacture. This continuum exploits all the processes found in a manufacturing plant.[16]

In large manufacturing plants, direct emissions and small spills attract the most environmentally focused attention. In practice, emissions and small spills are really indistinguishable, and each basic industrial process in a manufacturing plant offers different opportunities for them to occur. Maintenance processes that are commonly observed in manufacturing plants include[17]

- Forging and shaping

- Joining, soldering, and welding

- Cutting, drilling, grinding, and polishing

---

[16]This may explain why very little is written about the environmental concerns of maintenance per se. This discussion draws on the extensive literature on the environmental concerns of manufacturing, particularly the manufacture of aircraft and aircraft systems, large industrial vehicles such as farm and construction equipment, and sophisticated integrated electronics systems.

[17]The EPA's Office of Compliance has developed a series of notebooks that provide useful information on each of the processes shown here and detailed environmental concerns associated with them. The discussion here draws on U.S. Environmental Protection Agency, 1992, 1995a, 1995b; and case studies described in Berube et al., 1992; Breton et al., 1991; DeLange, 1994; Duke, 1994; Grogan, 1996; Lagoe, 1995/96; Loren, 1996; Marchetti et al., 1995/96; Maxwell et al., 1993; Ochsner et al., 1995/96; Owen, 1995b; Rank Xerox, 1995; and Wever, 1996a.

- Cleaning and etching

- Plating

- Applying paints, adhesives, and other coatings

- Testing and inspecting

- On-site recoveriy, recycling, and treatment of hazardous materials

- Storing and handling of material, including inventory in process

- Heating, ventilating, and air conditioning

- Lighting

Environmental concerns vary by the specific process used in a plant, but a variety of generic opportunities for emissions exist:

- Tanks of solvents, acids, and alkaline baths can produce air emissions or spills unless they are carefully sealed. Where vapors from these tanks are integral to production, vapor closure and recovery is important. So is careful management of vapor generation.

- Dragout from tanks of solvents and other chemicals can allow air emissions and can leave hazardous chemicals in the water used to rinse items after treatment.

- Contaminated baths must ultimately be removed and either recycled or treated and reduced to sludge or wastewater for disposal. Emissions can occur in each step of the management of these baths.

- Overspray during the application of chemicals creates a strong potential for emissions in the absence of proper closures and chemical capture systems. Overspray also contaminates tools and other industrial equipment that, unless it is disposable, must be cleaned, generating hazardous waste in sludge, wastewater, and potentially mixed solvents associated with the spraying and cleaning. Such hazardous waste presents the risk of emissions.

- Chemicals used as an integral part of the manufacturing process can escape during production in the absence of proper closures and capture. They can also escape following production while an item is in storage or even in use.

- Large-scale storage tanks and barrels of chemicals leak. Transfer of chemicals into smaller containers for use on the production line allows leaks, spills, and uncontrolled loss on the production line. A tendency to take more from supply than is needed for a particular maintenance job in order to simplify the work increases the potential for emissions.

- Dust and particulate matter from industrial operations can foul the workplace and leave the plant via air and wastewater if it is not managed properly.

- Faulty operation of filters, incinerators, and other pollution-control devices can allow emissions to escape.

- On-site boilers and electricity generators offer the potential for traditional air pollution and can generate ash that contains hazardous constituents.

- Chillers now use lower-risk ozone depleters, ammonia, and other substitutes for CFCs that all pose dangers in the workplace and all have the potential for leaks.

This far-from-complete list of concerns illustrates how pervasive chemicals are in a modern manufacturing facility and reminds us that no magic bullet exists to manage their emissions. One policy that touches several of these concerns is the use of pharmacy programs that closely regulate access to hazardous chemicals in a maintenance facility. Under a pharmacy plan, a central supply point issues only the amount required for a specific maintenance action at the time it is needed, to reduce the chance for uncontrolled chemical losses on the shop floor. More generally, many specific challenges are presented throughout the plant, challenges that must be met by each worker associated with processes that involve chemicals. Large central maintenance facilities in the DoD pose every one of these challenges. Responses to them can clearly benefit from the expertise of engineers who understand the production process and alternatives for managing environmental problems associated with it. The best results occur when engineers and production workers work well together.

All of the chemicals that do not leave the plant as emissions or in the products themselves ultimately appear as waste, often hazardous. The RCRA imposes serious oversight on the management and disposal of such waste. Liability for personal injuries, property and natural resource damage, and remediation of sites creates the potential for additional large, uncertain disposal-associated costs, even when an organization fully complies with RCRA and comparable state laws. Companies have sought to increase control over their wastes, limiting the number of sites they deal with or retaining the wastes to manage themselves. Ideally, firms seek to avoid disposal altogether. Waste exchanges offer opportunities for marketing waste to other firms that can use it as an input, but only limited use is made of such exchanges. More often, firms redesign their processes to recycle hazardous chemicals on site or, increasingly, to eliminate dependence on the chemicals altogether.

Nonhazardous waste, liquid and solid, also imposes costs, but they are not nearly as severe as those for hazardous wastes. Firms often treat hazardous waste sufficiently to allow its disposal as nonhazardous waste, but they are increasingly using techniques such as those discussed above with regard to packaging. In fact, packaging remains a major source of nonhazardous waste in manufacturing facilities. Firms are also building water treatment facilities on site to manage costs associated with industrial wastewater treatment. However, cost savings are limited by regulations that cover treatment activities on site, and firms often find that regulation is just as strict for on-site treatment as it is for treatment away from the plant.

Serious spills and fires pose a level of danger in manufacturing plants comparable to the danger they pose in warehouses. Many large plants maintain large storage tanks, and the manipulation of chemicals in the production process may actually increase the danger of serious accidents. The type of manufacturing analogous to central maintenance in the DoD does not typically use large enough quantities of chemicals to pose risks as high as those associated with large tanks, but nonetheless, emergency plans like those discussed above for warehouses are important here as well.

Manufacturing facilities are major energy consumers, and as noted earlier, energy use raises questions about the depletion of nonrenewable resources, especially in Europe. It also raises concerns about traditional air pollutants and, again especially in Europe, carbon dioxide emissions. The European context is obviously important to U.S. defense forces operating there and to U.S. efforts to coordinate policy with European allies. Appropriate responses in manufacturing plants are similar to those discussed above.

Control

Each part of the DoD handles the control function relevant to logistics differently. As we define control here, it includes

- Efforts in financial management to measure performance, track funding, and manage the information systems that do this.

- Efforts in plans and programs to build the budget and allocate funds to environmentally related activities.

- Efforts in specific functional and line offices responsible for implementing environmental policy and defending requests for funds to ensure compliance and to support pollution-prevention activities.

The next chapter focuses on lessons learned in commercial firms relevant to these activities; here, we briefly introduce the major issues to place them in the overall context of central logistics.

Even in organizations with proactive environmental management practices, the primary environmental concern of the control function is whether logistics functions are in compliance with the law at every location.[18] Environmental audits began as a systematic attempt to track progress on compliance in a complex regulatory environment. Very few industrial activities can report an absence of notices of violation (NOVs) from their regulators, and most organizations view these as an indicator that production processes and end-of-pipe controls must be tightened. Some accept NOVs as one consequence of an active relationship with a regulator and see a total absence of NOVs as a signal that their organization is not dealing aggressively enough with the regulators. Firms holding this position are among some of the most progressive in their general positions on environ-

_____

[18]See Lent and Wells, 1992, p. 380.

mental responsibility. Long-term mutual trust between company and regulator is important, and long-term agreement on what effective compliance means is integral to that trust.

To the extent that an industrial activity is clearly not in compliance, control is the function that must devise a plan for achieving compliance. Once compliance is in hand, or NOVs have reached an acceptably low level, control can be focused on future regulations and creating plans to meet them. Only when these concerns are satisfied can control turn to more proactive concerns of pollution prevention beyond the level required by regulation. Of course, pollution prevention can provide a method for reducing the number of NOVs and hence promoting compliance.

Increasingly, progressive firms justify their proactive environmental policies precisely as a way to stay far enough ahead of regulatory change so that they are not constantly struggling to achieve compliance. Such a stance gives these firms flexibility and time to achieve compliance on their own terms, typically at lower cost to their core concerns than a day-to-day struggle with the regulator would allow. Even these firms, however, have regulation constantly in the background and recognize that their first responsibility is to ensure a satisfactory level of compliance.

The second major environmental concern of control is resource allocation. How much money should the organization program for environmental activities and how should it allocate this money among such activities? These questions obviously raise other questions about what an "environmental activity" is and who actually controls the resources that fund it.

In the past, industrial firms typically treated the cost of the environmental staff, much pollution-control equipment, and any fines, penalties, or legal fees as centrally funded corporate overhead. All investments required for compliance were funded without much debate. Pollution-prevention investment, however, had to compete with other corporate investments, typically at a disadvantageous hurdle rate.

Increasingly, industrial firms want to associate environmental costs with the products that precipitate them. Business units must program for these costs and recover them from the revenues their products generate in the market. Compliance is mandatory, but it is important for business managers to find ways to comply cost-effectively. Pollution prevention should continue to compete for funding, but decisionmakers should carefully evaluate all the corporate costs that successful pollution prevention can eliminate.

Industrial firms at the cutting edge now address these questions strategically, asking how they can translate environmental concerns into opportunities to achieve strategic advantage relative to competitors (with similar environmental concerns).[19] This strategic perspective raises environmental resource issues in the corporate planning process that firms use to address the choice and design of future products and processes and the location of future facilities. Such high-level assessments require effective integration of environmental and core business concerns to ensure that environmental costs and benefits relevant to products and processes are properly represented in the detailed assessments that underlie strategic decisions.

Strategic and tactical decisions that treat environmental concerns as central to the organization's activities raise the third major concern of control: adequate information on the effect that environmental activities have on core business concerns and on the organization's progress in integrating them.

Ideally, proactive industrial firms seek variations on activity-based costing (ABC) to express environmental concerns in the language that governs core business activities—accounting. These variations on ABC seek to link *all* the costs associated with a product or process to that product or process—production cost; cost of pollution-control equipment; cost of administrative work, paperwork, and information systems required to manage the relationship with regulators and other external stakeholders; cost of penalties, fines, and other legal expenses; cost of management focus within the firm; cost associated with lost market share resulting from adverse consumer reactions to corporate actions and products. No one has a cost-accounting system in place to do this on a routine basis. Advocates of ABC in general recognize the difficulties of changing whole accounting systems and recommend interim steps to apply ABC to

---

[19]Blumenfeld and Montrone, 1995; Porter and van der Linde, 1995; Wever, 1996b.

existing data sources. Proactive firms are doing this today with increasing success.

To track information on the performance of environmental management within a firm, innovative industrial firms increasingly turn to the TQM-based systems they use to track general corporate performance. There are many variations on these systems, which are discussed in greater detail in Chapter Three.

CONTRACTING

The foregoing discussion has focused on environmental concerns relevant to organic provision of central logistics services. In fact, the DoD contracts for a significant portion of transportation and maintenance services and is currently planning to expand its use of external sources for all central logistics functions. The issues above all arise in the context of contractor sources of central logistics functions, but they take on a somewhat different meaning for the DoD when they arise in a contractor setting.

Commercial firms worry about the environmental performance of their suppliers for two different reasons. First, the products they buy can affect their own environmental performance; they monitor the performance of their suppliers to limit effects on themselves when they use the suppliers' products. Second, integrated supply chain management does not respect organizational lines. It considers every aspect of the inputs and outputs of a process in search of opportunities to eliminate activities that do not contribute value for the final customer. Firms seeking to project an environmentally responsible image are uncomfortable buying products from firms lower in the supply chain that they know to be less than environmentally responsible. One may wonder if they have exploited integrated supply chain management to "export" their environmental problems to another firm. The DoD faces both problems with its own suppliers.

Practices differ across organizations, partly because of the relative importance of these two considerations and partly because of differences in production processes and organizational cultures. Firms that use TQM also apply quality templates to suppliers. Indeed, the

ISO 9000 series has become the standard tool that quality-based firms use to determine the quality of their suppliers' processes,[20] and the ISO 14000 series may come to provide a similar service with regard to environmental management systems.[21]   Other firms require their suppliers to adopt environmental management systems comparable to their own. In the future, ISO 14000 may provide the common ground that all firms use to confirm their joint commitment to environmental responsibility.

Firms concerned only about how a supplier's products affect their own environmental performance can demand detailed information on the chemical content of the products, as long as they make arrangements to protect proprietary data when they are delivered. These firms can also write contracts to encourage suppliers to make their products more environmentally friendly, e.g., they can include award fees and metrics to do this. This latter approach gives a buyer much broader latitude to address questions about packaging that generates nonhazardous waste or maintenance concepts that require the buyer to use hazardous materials to support a product that it has bought, even if the product itself contains no hazardous materials.

Responsible suppliers have incentives of their own not to sell products that involve hazardous materials. They may also have concerns about their environmental images. Or they may fear being held liable for damages caused by their products, even if users do not use the products properly. Product liability laws give plaintiffs broad latitude to recover damages from a product supplier, even if the product is not properly used or if its contribution to an injury is

---

[20]The ISO 9000 series requires suppliers to have a third party audit them periodically and certify that their management processes meet specifically defined standards. At its heart are three "quality systems," or "contractual models," approved in 1987. Of the series, 9001 is the most comprehensive and covers design and development, production, installation, and servicing of products; 9002 covers production and installation of products; and 9003 covers final inspection and test of products. Other guidelines explain the auditing approach itself and define how audits will occur. Many good references are available. See, for example, Johnson, 1993.

[21]The first element of ISO 14000, 14001, was approved for use in 1996. An active debate is currently under way about how organizations will in fact use ISO 14001 and the forthcoming parts of the series. Supporters see the potential for ultimately integrating the 9000 and 14000 series to implement TQM throughout an organization; doubters worry that ISO 14000 does not require a third-party auditor or an active pollution-prevention program. See, for example, Begley, 1996, and Butner, 1996.

minor. In the most extreme reaction to this concern, some chemical firms now retain ownership of their most hazardous chemicals and essentially rent them for use, under the close scrutiny of the supplier. Others use their own hazardous products at their own sites, under their own supervision, to provide services that buyers formerly produced themselves, using the supplier's product. Buyers can take advantage of these services to export their own environmental concerns, but those who are worried about the integrated supply chain must still monitor the production of such services.

SUMMARY

Central logistics in the DoD includes transportation, supply, maintenance, and control activities at the depot level. It includes both organic and contractor depot-level services.

Innovative organizations in both the DoD and the private sector are using an integrated supply chain perspective to integrate their logistics activities with one another and with their core organizational missions. This perspective looks at all the processes in the supply chain from the customer's point of view and asks how they can be improved to yield greater value to the customer. Properly understood, this perspective is compatible with the dominant idea in innovative environmental management today—integrating decisions on corporate environmental policy and its implementation with the core decisions of the organization. In fact, this approach to environmental management can be seen as a way to ensure that environmental concerns are appropriately addressed in an integrated supply chain.

A review of the environmental concerns that arise in logistics functions reveals a rich variety of issues. Transportation and supply issues focus on fuel consumption and packaging but raise many other concerns. Maintenance raises almost all of the issues that arise in manufacturing, from direct emissions to hazardous-waste generation. These issues arise in organic and contract settings, but when an organization contracts for logistics services, it must develop a clear understanding of which aspects of its suppliers' environmental performance are relevant.

No one technical solution can have a large effect on any of these functions. Transporters and supply specialists have many levers, each of which can affect a portion of the fuel-consumption problem. Packaging must be addressed for one product class, supplier, or customer at a time. Specific environmental issues arise in every part of the maintenance process. Taken together, these issues present many opportunities for improvement. To make the most of these opportunities, an organization needs a system view that links the effects of options to the final customer. It also needs motivation and capabilities throughout the system to induce people to pursue these options.

The environmental challenge to central logistics is that of finding an approach to environmental management that allows logisticians to pursue this rich set of opportunities cost-effectively. The discussion in this chapter hints repeatedly at answers to this challenge. The next chapter confronts the challenge directly.

Chapter Three

APPROACHES TO ENVIRONMENTAL MANAGEMENT IN PROACTIVE COMMERCIAL FIRMS

Corporate environmental stewardship. Environmentally sustainable industrial practice. Eco-efficiency. Responsible care. Corporate public relations hype from companies concerned about their environmental image? Or do these terms represent a reexamination of corporate strategies and behavior? In some cases it may be more talk than action. Over the last two decades, however, environmental considerations have become an increasingly important aspect of doing business. As a result many companies have adopted aggressive environmental initiatives.[1]

During the past 20 years, a remarkable consensus has emerged among innovative commercial firms that support an increasingly proactive approach to environmental management. The details are still evolving and differ from one firm to another, but in broad outline, a common approach is emerging.[2]

This chapter first discusses why these firms see a need for change and what barriers they are experiencing. It then summarizes several recent surveys of commercial experience to demonstrate that although the surveys emphasize different elements of change, they are all consistent with one another.

The discussion then turns to a more detailed description of how proactive firms are approaching environmental management policy.

---

[1]Richards and Frosch, 1994, p. 1.

[2]For a recent overview of the academic literature and several key issues relevant to this approach, see Starik et al., 2000.

It argues that the principle underlying most of this change is that an organization must integrate its environmental decisionmaking into its core decisionmaking process.

Once a policy is developed, an organization must implement it. Proactive firms that are implementing their environmental management policies are placing heavy reliance on TQM. Many firms rely on a homegrown approach to quality; many others take advantage of formal quality templates to ensure that they are in fact implementing TQM.

The chapter closes with a brief overview of the consensus, derived from a quality perspective, that is developing among proactive firms.

MOTIVATION FOR CHANGE

The past two decades have seen unprecedented expansion of environmental regulation around the world. In the United States alone, firms spend $140 billion a year to improve their environmental performance, and spending is continuing to grow by 9 percent annually.[3] Texaco, for example, over a five-year period, expects to spend three times as much on compliance and emission reductions as the firm is worth in book value.[4] This case is exceptional, but it is indicative of the challenge that large industrial firms face today. For many years, firms thought that they could resist the growth of environmental regulation. They were wrong.

Today, the regulatory environment remains uncertain and threatening to any firm relying heavily on chemicals. The Bhopal disaster in 1984 demonstrated how badly things can go in the presence of modern industrial chemicals.[5] Uncertainty affects even industrial activities once thought to be perfectly safe. In 1987, the Montreal Protocol banned a class of industrial chemicals worldwide that only 15 years earlier were believed to be among the safest in use.[6] Firms

---

[3]Owen, 1995b, p. 59.

[4]Walley and Whitehead, 1994, p. 46.

[5]Piasecki, 1995, describes the corporate reaction to Bhopal.

[6]Palmer, 1980, documents the implications of banning stratospheric ozone depleters.

are certain only that the regulatory environment will continue to shift and that they must find ways to work in this environment.

On a less dramatic level, many corporate executives were surprised when they saw the Toxic Release Inventory (TRI) data first released in 1988.[7] These data document each year the amount of major hazardous chemicals large firms emitted or transferred from their plants in the previous year. Executives skeptical about environmental regulation in the past realized that the public would not tolerate such releases in the future. Since 1988, TRI data have become a prime focus of every large corporation that issues an annual public report on its environmental performance.

In this environment, the majority of industrial firms remain content to focus on complying with current law.[8] But a growing number of firms see opportunities in confronting the high and uncertain cost of environmental regulation proactively. Some actively seek opportunities to reduce their costs through pollution prevention or negotiation with regulators to find more cost-effective ways to comply. Others see strategic advantage in the shifting regulatory environment and are developing ways to improve their competitiveness through better environmental management.[9]

The reasons for seeking to reduce cost are self-evident: Environmental compliance imposes a growing cost over time precisely when firms are under severe competitive stress to improve their processes and cut long-term costs. Proactive firms differ mainly over the size of the potential for cost savings. Some see dramatic opportunities; others fear too much attention given to cutting costs of pollution prevention will divert the firm from more strategically important problems.[10]

---

[7]Case studies make this point repeatedly. See, for example, Megna and Savoy, 1991, p. 411; Ochsner et al., 1995/96, p. 66; Owen, 1995b, p. 60.

[8]Results of a 1991 survey by Abt Associates reported in Clarke et al., 1994. Other recent surveys yield similar results. See, for example, the A. D. Little survey results reported in Kirschner, 1996, p. 19.

[9]For a useful exchange of views on these alternative perspectives, see Walley and Whitehead, 1994; Clarke et al., 1994; and Porter and van der Linde, 1995. See also Lent and Wells, 1992; and Ehrenfeld and Howard, 1995.

[10]Clarke et al., 1994, offers a good overview of this range of views.

The strategic view is more subtle. Some consumer firms see an emerging market for "green" products or products sold by "green" companies. To date, actual behavior has not borne out consumer claims in surveys that they would pay a premium for a green product, but firms remain hopeful that faced with two products that cost the same, consumers will prefer the greener product.[11] Industrial firms, of course, seek to buy green products for different reasons, as discussed in Chapter Two.

Strategists, however, seek to be proactive for reasons other than simply designing new products. Many firms claim that it is easier to manage environmental costs if they have time to do research and development (R&D) and make investments relevant to pollution prevention.[12] Others point to the complexity of environmental management in a global setting. It is easier to manage a uniform corporate policy worldwide than to customize the corporate policy to every locale. Corporate policy should look beyond specific regulations—for example, anticipating the migration of regulations from one part of the globe to another—and let local business units manage the specifics of compliance.[13] For still other firms, the strategic concern is more basic. According to Charles Bravo of the U.S. Postal Service, "If you can move your staff away . . . from the use of certain chemicals that are hazardous, then you can be freed from the laws governing those chemicals. Not only do you reduce your potential liabilities, but you also free up staff time to focus on performance improvement and business opportunities, not just compliance."[14] Proactive environmental management is a way for an organization to maintain its strategic focus on its core competencies.

In sum, proactive firms have many different motives, and their specific actions reflect these differences. But a growing number of firms believe that proactive environmental management makes good business sense.

---

[11]Blumenfeld and Montrone, 1995, p. 83; Breton et al., 1991, p. 390; Maxwell et al., 1993, p. 321.

[12]See, for example, Penman and Stock, 1994, p. 849; Rank Xerox, 1995, p. 20.

[13]Blumenfeld and Montrone, 1995, p. 86; Breton et al., 1991, p. 390.

[14]Quoted in Blumenfeld and Montrone, 1995, p. 88.

BARRIERS TO CHANGE

Organizational change is not easy. The literature indicates that those who want to change have encountered economic, organizational, and regulatory barriers.[15]

Three kinds of *economic* barriers are important. First, firms limit the funds available to invest in pollution prevention and other proactive environmental activities. They focus their resources on core activities and see environmental issues as ancillary to the main concerns of the business. Limitation occurs when resources are not committed to the staff required to set up and execute proactive programs and when pollution-prevention investments must achieve a higher hurdle rate than other investments to be approved.

Second, the personnel in the organization who might identify opportunities to cut environmental costs see no benefit from cutting those costs. When a firm budgets for pollution control from an overhead account, a local manager cannot increase his or her business unit's profits by cutting those costs. Conversely, when accountants treat pollution-prevention expenditures as liabilities rather than assets, pollution prevention hurts the books, wherever they are kept in the firm.[16] And as often as not, bonuses and advancement opportunities depend more on improving core corporate activities than on improving environmental performance.

Third, the benefits of proactive environmental actions are difficult to quantify. Cost savings typically accrue to accounts in the firm not directly linked to a production process where an investment might be made. That makes it hard to track costs, even if the firm wants to. Strategic benefits of the type discussed above are not generally quantified. Managers must use judgment to weigh their importance, making the benefits of environmental investments less tangible than those of simpler production investments.

---

[15]For a particularly good overview, see Marchetti et al., 1995/96. Sources differ about whether particular barriers are economic, organizational, or regulatory. The classification offered here is as useful as any other in the literature.

[16]Hoffman, 1992/93, p. 6.

Four types of *organizational* barriers are important. First, it is well documented that large organizations naturally resist change. The changes a firm makes to become more proactive typically disrupt the organization by forcing new paths of information flow and overriding standard operating procedures. Previous RAND analyses have shown, for example, that the natural inertia of these procedures and traditional flows of information can simply choke a new program if there are not sufficient implementation levers in place to realize the innovators' intent.[17] And if changes actually threaten turf in an organization, active opposition can be expected. Proactive firms find that opposition often comes from within the environmental management community itself—from personnel responsible for pollution control and other compliance activities who see their jobs threatened by pollution prevention.[18]

Second, personnel in a firm may believe that proactive policies threaten their performance, if not their jobs. Managers who have made major investments in equipment that pollution-prevention actions could alter or replace see the return on those investments threatened.[19] Managers and workers responsible for making the production line work day to day resist the efforts to experiment with the line that are required to design and implement many forms of pollution prevention.[20] Managers especially resist untested technologies—including many pollution-prevention techniques—that could threaten their ability to generate product from the production process. For example, many managers have a natural aversion to using recycled materials, which they consider inherently inferior to new materials, even if they offer as much as a 25 percent discount.[21]

Third, corporate emphasis on short-term performance can discourage investment. Bonuses and advancement opportunities often depend on this year's net income or even this quarter's net income. In such a setting, proactive environmental activities impose costs dur-

---

[17]See Camm, 1996, for a discussion of this problem and further references.

[18]Hoffman, 1992/93, p. 5.

[19]Richards and Frosch, 1994, p. 3.

[20]Berube et al., 1992, p. 193; Ochsner et al., 1995/96, pp. 71–72.

[21]Penman and Stock, 1994, p. 842; Richards and Frosch, 1994, p. 3. In the DoD, of course, military specifications typically support this bias against recycled materials.

ing the review period and typically generate a meaningful payback only after the review period is over.

Finally, organizations fail to give proactive environmental activities enough support. The senior leadership is not involved in these activities or vocal about the importance it places on them. Managers of business units, plants, or shops provide inadequate support and focus. Such lack of active line support leaves environmental specialists unempowered. It also leaves specific responsibilities unassigned, so individuals do not know that they are expected to be and will be held accountable for being proactive.

Two kinds of *regulatory* barriers appear to be important. The first is similar to the resistance to change experienced within a regulated firm. Regulators responsible for permitting end-of-the-pipe oversight are threatened by innovative environmental policies that might do away with the need for traditional permits or end-of-the-pipe regulation. They may feel a threat to their jobs or, failing that, a threat to their effectiveness as regulators. This second concern is especially true of young regulators, who lack the experience of their corporate counterparts in environmental management. These regulators may fear that regulatory change giving firms more discretion shifts the balance of power still more in the direction of their more experienced and skilled counterparts.

The second kind of regulatory barrier derives from the regulation of treatment and recycling activities on site. The RCRA and many state agencies maintain close controls over these activities,[22] often with the result that a firm pursues a treatment or recycling option in the hope of escaping some regulatory burden only to discover that it has traded one regulatory regime for another. Some firms fear that the regulators use control over on-site activities precisely to avoid being displaced by pollution-prevention and other innovative environmental policies.

Individual firms experience these barriers in varying degrees, and their approaches to innovative environmental management clearly

---

[22]See, for example, Duke, 1994, pp. 453–455; and Richards and Frosch, 1994, p. 3. Berube et al., 1992, provides an especially useful case study in which this problem plays a central part.

reflect the barriers they have encountered. Because all of these barriers appear potentially relevant to the DoD, it should find an amalgam of firms' efforts to breach them useful.

THE BASIC CONSENSUS ON A PROACTIVE APPROACH

As efforts to deal with barriers have moved ahead, several groups have surveyed corporate environmental management programs and summarized the features that appeared to be most important to their success. Different surveys have yielded broadly compatible outcomes. We offer three here to illustrate the general way in which specialists in environmental management now think about best practices.[23]

The first is a joint effort in 1993 by AT&T and Intel to benchmark pollution prevention at five firms that they determined were among the best in the world at it: Dow, DuPont, H. B. Fuller, 3M, and Xerox. Pollution prevention is only one element of environmental management, but it poses most, if not all, of the problems that an effective, proactive environmental management program must address.[24] AT&T and Intel used AT&T's standard benchmarking technique to conduct the study.[25] Table 3.1 summarizes their findings.

The activities identified as "essential" or "critical" to the success of deploying and then sustaining a pollution-prevention program address precisely the kinds of barriers discussed above. Top-level commitment is necessary to verify the lasting importance of pollution prevention. The leadership must support the effort from the beginning and remain informed about its progress, and a formal program must be set up to take day-to-day responsibility for pollution-prevention policy.

---

[23]The approaches here are compatible with total quality approaches to environmental management. We discuss them later, after we have reviewed information on successful environmental management approaches in general.

[24]Pollution prevention has also done the most to induce firms to manage environmental issues proactively. See Lent and Wells, 1992, for the results of a survey.

[25]For details on the approach, see Klafter et al., 1993.

Table 3.1

Key Elements in a Pollution-Prevention Program, Based on AT&T/Intel Benchmarking Study[26]

Topic	Critical or Essential to *Deploying* a Best-in-Class Pollution-Prevention Program	Critical or Essential to *Sustaining* a Best-in-Class Pollution-Prevention Program
Structure	Top-level commitment Corporate EHS organization Dedicated pollution-prevention staff Feedback to CEO	Top-level commitment Corporate EHS organization Dedicated pollution-prevention staff Cross-functional teams Feedback to CEO
Framework		Institutionalized policy/goals/procedures Strategic objectives for pollution prevention in business plans Formal process (standards and practices)
Strategic direction	Mission Goals Policy Procedure	
Selling the program	Communication of strategic direction Description of pollution-prevention program Secure buy-in	
Motivation	Top-management encouragement	Pollution-prevention empowerment encouraged Pollution-prevention reward program Top-management recognition Tied to compensation Tied to performance appraisal
Research		R&D support for pollution prevention Pollution-prevention R&D integrated into process, design, and manufacturing equipment design
Communication		Internal technology transfer Internal communications media

Source: Klafter et al., 1993, pp. 27–29.
CEO = chief executive officer; EHS = environmental, health, and safety.

---

[26]These results are compatible with those of a similar study by the Business Roundtable, using the AT&T approach, of facilities owned by DuPont, Intel, Martin Marietta, Monsanto, Proctor and Gamble, and 3M. See Business Roundtable, 1993.

To overcome organizational inertia, the leadership must be clear from the start about the mission, goals, policies, and procedures associated with the new program. The pollution-prevention program must then communicate this information throughout the organization and secure the buy-in of the line managers of the relevant business units and plants. At the beginning, selling the program will rely heavily on encouragement from the senior leadership.

Over time, the pollution-prevention program should set up cross-functional teams to link environmental with core production process personnel. These teams can help develop a fully institutionalized process for generating goals and objectives and integrating them with the core business processes of the firm. The motivation program should be formalized to ensure that pollution prevention is integrated into all appropriate incentive programs in the firm. As specific needs can be identified, a formal R&D program for pollution prevention should be initiated, and, more important, pollution prevention should be encouraged as an active element of R&D conducted on the core activities of the firm. The pollution-prevention program will support learning throughout the firm and communicate information about successes around the firm to build momentum for the program.

In 1994, the National Academy of Engineering published the results of a conference on corporate environmental practices that yielded very similar results. It identified the following "emerging principles of environmental practice":[27]

- Search systematically for ways to reduce adverse environmental impacts.

- Provide leadership and demonstrate a commitment to minimizing environmental impacts and improving the environment.

- Ensure broad corporate involvement in environmental initiatives.

- Encourage and provide incentives for innovation.

- Establish open communication with all stakeholders.

---

[27]Richards and Frosch, 1994, p. 2.

- Develop clear targets and effective measurement systems.

- Respond to, do not be paralyzed by, uncertainty.

These points are completely compatible with the principles identified by AT&T and Intel, although they provide a somewhat different emphasis and identify a few new ideas. The National Academy confirms the importance of leadership, a proactive stance, clear goals, and clear incentives. It adds the importance of quantitative information systems and the need to communicate with all stakeholders, not only those inside the firm. And the National Academy also recognizes a key problem—uncertainty—that often arises as a barrier when a firm asks its employees to do something new. The principles encourage them to be proactive, not conservative, when they cannot easily predict the results of their actions; implicitly, firms are told to create incentives that encourage such behavior.

In 1996, Robert Shelton, director of environmental, health, and safety consulting at A. D. Little, advised environmental managers to adopt five principles as a basis for a more proactive stance:[28]

- Develop measures of success that show business contributions.

- Use business frameworks—look at "competitive advantages."

- Develop a basis for collaboration, not command and control.

- Use business language, not "environmentalese."

- Align environmental, health, and safety thinking with the business perspective and the drive to add value.

Shelton's proposed approach is, again, thoroughly consistent with those of AT&T and Intel and the National Academy of Engineering. But again, Shelton changes the emphasis. He encourages environmental managers to move more closely to the core of the firm, saying that they should seek environmental improvements that, when judged by standard business processes, clearly contribute to the business itself. He recognizes the importance of institutionalizing a successful environmental program by integrating it with core pro-

---

[28]Quoted in Kirschner, 1996, p. 19.

cesses. And he places the onus on environmental managers to do what is necessary to make this integration work.

These different perspectives on a common problem show how different commentators have emphasized different aspects of the total picture.[29] In the following, we examine environmental management in greater detail to build a total picture of the potential offered by innovative practice from the individual pieces observed in different cases and settings.[30]

DEFINING A PROACTIVE ENVIRONMENTAL MANAGEMENT PROGRAM

An effective environmental program is simple in principle. It starts by identifying relevant stakeholders, since anything a program tries to do must be tied to a stakeholder. A proactive firm does this by identifying each stakeholder's preferences and translating them into goals for the firm. The firm then translates these goals into concrete objectives and metrics that can be used to track progress. The firm must have an information system to track its performance, and its metrics must support the resource management system that the firm uses to decide where to deploy its resources and what investments to make. The firm needs an effective costing system to do this. Finally, the firm needs to develop a relationship with its suppliers that allows it to make effective resource management decisions about them as well as about its organic activities.

In practice, the success of these efforts hinges on the firm's ability to integrate its environmental activities with its core business activities, as well as on critical implementation issues associated with effective leadership, empowerment, motivation, and issues of institutionalization.

---

[29]The U.S. Environmental Protection Agency (1994, 1996a, 1996b) covers many of the same points in the context of managing a pollution-prevention program.

[30]Ochsner et al., 1995/96, provides an exceptionally useful case that illustrates many of the points discussed here and shows how these points fit together in the context of one case. We frequently cite Ochsner et al., 1995/96, with respect to specific issues as we review them.

Almost everything said about a firm can apply—often quite directly—to the DoD. The DoD has its own stakeholders; goals, objectives, and metrics relevant to each of them; and information systems. Its central logistics supply chain includes both major organic and contractor activities.

Stakeholders, Goals, Objectives, and Metrics

Innovative firms typically identify basic corporate goals that reflect the needs of their stakeholders and help the firm trade off the interests of those stakeholders. The goals are translated into more concrete objectives that evolve over time to reflect progress on environmental management and changes in the business environment relevant to environmental management. The firms develop systems of quantitative and qualitative metrics to track their progress against their objectives.

Shareholders are often explicitly identified in the annual reports of innovative firms. The opening sentences of the John Deere Vision Statement are typical:

> The purpose of Deere & Company is to create value for all of our constituents. Our constituents include customers, employees, shareholders, and the global community. Our customers have invested in our products; our employees have invested their energy and creativity; our shareholders have invested capital; and the global community has invested the use of its environment.[31]

Goals and priorities flow from open and continuing communication with stakeholders. The goals must be broad enough to apply companywide and yet precise enough to provide useful guidance on how to make tradeoffs among the interests of different stakeholders. The most common environmentally oriented goals in the literature give first priority to eliminating immediate threats to human health and safety and the next priority to eliminating threats to the natural environment. Once these threats are under control, efforts can be undertaken to reduce costs or improve other elements of corporate performance.

---

[31] *Deere & Company Fact Book 1995*, available at www.deere.com.

DuPont's titanium dioxide business unit, for example, gives highest priority to activities required to eliminate safety problems or to comply with environmental regulations. Then it calls for activities that increase capacity or improve the quality of the product. After that, it considers activities that are clearly cost-effective under DuPont's normal project evaluation standards.[32] Rank Xerox clearly states preferences similar to those of the Environmental Protection Agency (EPA): Seek elimination of an effluent stream first, then substitution, minimization, control, and finally protection.[33] Weyerhaeuser, in its chemicals control program, has boiled a similar approach down to the idea that "less is more."[34] These priorities leave much room for interpretation, but they provide clear guidance on the concerns that must be addressed when weighing alternatives.

Goals also provide the basis for more specific objectives, i.e., those that target specific activities. For example, AT&T set six key environmental objectives in 1990:[35]

1. Phase out CFC emissions from manufacturing (50 percent by year-end 1991, 100 percent by year-end 1994).

2. Eliminate toxic air emissions (TRI) (50 percent by year-end 1991, 95 percent by year-end 1995, near 100 percent by year-end 2000).

3. Reduce manufacturing process waste disposal by 25 percent by year-end 1994.

4. Recycle 35 percent of internally used paper by year-end 1994.

5. Reduce internal paper use by 15 percent by year-end 1994.

6. Improve occupational safety: 100 percent of AT&T manufacturing facilities accept Occupational Health & Safety Administration (OSHA) Voluntary Protection Program by year-end 1995; 50 percent gain STAR status.

---

[32]Ochsner et al., 1995/96, p. 68.

[33]Rank Xerox, 1995, p. 18. The EPA management "hierarchy," defined in the Pollution Prevention Act of 1990, favors first source reduction, then recycling, treatment, and as a last resort, disposal.

[34]Loren, 1996, p. 105.

[35]Breton et al., 1991, pp. 390–391.

Such objectives target specific activities and have firm, quantitative targets linked to clear schedules. They lay the groundwork for an auditing program that can track progress toward achieving the stated objectives and hold groups accountable for progress.

AT&T's objectives raise three important points. First, who sets these objectives? Some companies set them at the corporate level, as AT&T did. More commonly, firms adopt general goals at the corporate level and allow their major business units to translate these into specific objectives.[36] Where such policies are created obviously depends on the prevailing organizational culture in the firm. When firms delegate the development of objectives, the corporate level typically uses competition and benchmarking among business units to motivate ambition, check for feasibility, and validate specific targets and deadlines.

Second, how much detailed analysis do firms conduct before they set high-level goals? Firms obviously differ, but most elicit information from the field and then, without knowing exactly how the targets will be met, use a top-down process to set them. Targets vary in ambition, but many firms encourage "stretch" goals that they may only have a low—say, 25 percent—chance of realizing.[37] The targets are meant to challenge everyone in a firm to think about the objectives differently and hence to break traditional mindsets. If the targets cannot be realized, the organizations learn from this discovery and continue to press for qualitative change in their processes.

AT&T, for example, discovered after the corporate headquarters identified these objectives that the business units could not baseline their actual paper use, which made it impossible to monitor the fifth objective above. They decided that a corporate program could best focus its attention on places where corporate policies *could* change behavior, e.g., data centers, training centers, and centers that manage forms, all of which consume large quantities of paper. AT&T developed effective baselines and monitoring mechanisms for the other goals and hence has succeeded in pursuing them more comprehensively.

---

[36]See Maxwell et al., 1993, p. 322

[37]Junkins, 1994, p. 58.

Third, what objectives should a firm target? AT&T's objectives are an eclectic mix. Only one, the first, is tied directly to compliance. The second seeks to respond to a regulation that requires public reporting of TRI data but no actual reduction. The others reflect AT&T's judgment that environmental and safety improvement is cost-effective for the firm. But why this particular set of objectives?

Corporate approaches to choosing environmental targets for emphasis vary dramatically. The Caterpillar plant in East Peoria, Illinois, started with "low-hanging fruit," which it identifies as toxic, high-volume, or high-cost waste streams that it can eliminate or reduce.[38] Ummenhofer recommends that an organization give an environmental objective priority commensurate with its relevance to the operation of logistics and its relevance today as opposed to the foreseeable or potential future.[39] Bailey and Soyka, on the other hand, think of low-hanging fruit from a different perspective:[40] Their analysis suggests that a firm should focus its environmental management concerns first on activities distant from the corporate core concerns—that is, where business risks are lower, less cross-functional and senior management involvement is needed, and decisions are easier to reverse. From this perspective, a firm should focus on pollution-prevention activities, utilities management, basic facilities maintenance procedures, and waste management before turning its attention to issues central to the production process or product definition.

These perspectives illustrate the competing demands on environmental managers. They suggest that corporate goals should be set to support a long-term program of improvement and that objectives should be chosen and revised on a continuing basis. At the beginning of an environmental management program, objectives should reflect immediate compliance concerns, but seeds should also be sown for a longer-term program of proactive management. This longer-term program should start, where possible, with lower-visibility activities that the firm can pursue without major changes in its current culture. It should choose these activities in anticipation of

---

[38]Owen, 1995b, p. 59.

[39]Ummenhofer, 1995, p. 28.

[40]Bailey and Soyka, 1996b, pp. 15–20

more ambitious future objectives. Hence, the firm should seek to introduce long-term goals in each business unit or plant and in each generic activity to begin building experience with and support for a broader program. As it realizes success in these activities, the firm can in fact build a case for change and use its growing experience to take on more challenging initiatives.

The firm's objectives act as its vehicle to communicate decisions about its specific priorities as its program proceeds. As the environmental management program grows, its objectives become an increasingly integral part of the general business plan of the firm.[41]

To track its success against its objectives, a firm needs specific metrics that are clear and simple.[42] The AT&T example above focused on quantitative targets, suggesting a strong preference for quantitative metrics. But most firms find that other metrics can be useful as well. In particular, the dynamic program of environmental management suggested above points to a formal process of management underlying the various goals and objectives the firm chooses.

Many firms adopt qualitative metrics to monitor this ongoing process.[43] For example, they may survey their key stakeholders on a regular basis and ask each of them to assess the firm's performance from its own point of view, using qualitative scales to capture different aspects of performance and their relationship to the stakeholder. They may also use regular qualitative self-assessment by employees involved in the processes relevant to water conservation, packaging, transportation, hazardous waste, and so on. Qualitative scales can help summarize impressions here, but simple verbal reports also help identify problem areas for immediate attention.

The objectives, and hence the metrics relevant to environmental management, differ at different levels in the firm.[44] At the very highest level, they reflect the broadest concerns of the firm and look very much like the long-term goals the firm uses to initiate its environ-

---

[41]See Klafter et al., 1993, p. 41.

[42]Haines, 1993, p. 367.

[43]Brown and Dray, 1996, pp. 71–75.

[44]Kaplan and Norton, 1996; see specific case studies in Kaplan, 1990.

mental management efforts. At each business unit and plant, the objectives become more specific to the activities in that locale. They give greater attention to the processes there, and they are phrased in terms relevant to the activities that employees can affect in these processes. In fact, at each level, a manager and a subordinate negotiate a specific version of the metrics that will be used, reflecting to the full extent possible the firm's broad goals and the specific factors that the subordinate can affect. Such negotiated metrics change over time as the manager and the subordinate renegotiate their relationship. This is true even if plant, business-unit, or corporate objectives remain stable over time. In sum, objectives and metrics do not simply roll down through an organization; they adjust to the decision environment at each locale within it.

Data on Environmental Performance

Using metrics to track performance implicitly assumes that data exist to support such metrics. Our survey of the field suggests that it may be most appropriate for the DoD to tailor metrics early in an environmental management program to the data available in existing systems or to estimate appropriate metric values based on these systems. But over time, accountability demands an auditable—internally or externally—information flow. In all likelihood, a new environmental management program will identify information needs that the firm will have to accommodate over time in its information architecture. Even in the short term, basic data on effluent levels are important for compliance and pollution-prevention planning.

The case of DuPont is worth noting, given its salience to DoD concerns. When DuPont decided in 1987 to reduce its emissions of carbonyl sulfide, one of its first steps was the introduction of new monitoring equipment at each of the affected plants.[45] The equipment alone cost $300,000 per plant. The real-time data gave DuPont immediate evidence that the changes it was testing had the intended effects. This immediate evidence helped environmental specialists work with process operators to assure them that the risks they were

---

[45]TRI data show that DuPont had more chemical emissions than any other source between 1987 and 1993. Carbonyl sulfide accounted for 30 percent of its air emissions in 1987 (Ochsner, 1995/96, pp. 66–67).

taking and the inconvenience they were experiencing were worth the costs. The changes DuPont made based on these data saved the firm $8 million to $14 million per plant in capital costs and $400,000 per year per plant in raw material costs. And the data system continues to support other improvement efforts of various kinds.

Innovative firms have recognized the importance of detailed information on environmental performance since at least the 1970s. At that time, multinational chemical and petroleum firms found that it was too difficult to comply with the regulations they faced without formal accounting systems.[46] They developed compliance audits that identified all relevant regulations and created complex checklists that could be used to verify compliance.

Such systems enabled the senior leadership to understand and appreciate the scope of their compliance problems. They also provided data sources for pioneering programs to reduce compliance problems by reducing emissions of regulated chemicals. As such efforts found success in individual firms, the appeal of a more proactive form of auditing grew.

Nonetheless, proactive compliance audits are still not typical, even in large firms. A 1995 Price Waterhouse survey of 445 firms found that only 73 used environmental audits.[47] That is a small number, but it reflects rapid growth—60 percent of these firms had initiated their audit programs within the previous three years. Nevertheless, only a fraction (unidentified) of them currently use proactive "self assessment and validation" audits rather than classic compliance audits to identify corrective actions.[48]

As firms become more comfortable with self-audits, they are beginning to think about how to link their environmental data systems with related systems used to manage material flows and preventive maintenance.[49] Such efforts are part of a broader interest in integrating environmental management into core management activi-

---

[46]See Willig, 1995, for information on these firms and their accounting systems.

[47]"The Green Machine," 1995, pp. 17–18.

[48]Rank Xerox, 1995, p. 7.

[49]Richards and Frosch, 1994, p. 23; Grogan, 1996, p. 57.

ties.[50] Such integration provides a basis for applying many of the improvement tools that these firms use to manipulate data from their core data systems. In principle, environmental concerns could become just one more point of focus for such improvement efforts.

Any discussion of data systems and information architectures in innovative firms ultimately finds its way back to using these systems to promote learning and improvement throughout the firm. These firms place a high premium on understanding current performance levels in different parts of the firm and diffusing information through the firm to promote improvement.[51] A high premium is also placed on tracking improvements outside the firm and seeking ways to bring information on external innovations into their own systems.[52] And to promote such behavior, the firms place a high premium on creating learning organizations that can benefit as much as possible from the information systems they have put in place.[53]

Justifying Environmental Management Decisions

Firms have little difficulty justifying expenditures required for compliance. Firms typically approve such expenditures without any financial analyses.[54] But the more proactive approaches to environmental management increasingly require firms to choose between discretionary environmental spending and spending on other activities. Not surprisingly, the innovative firms wish to apply the same criteria and cost-benefit standards to environmental and non-environmental spending.

In case after case identified in the literature, individual firms justify discretionary spending on one of two grounds. First, the expenditure is cost-effective on the basis of traditional project evaluation criteria. As noted above, DuPont could justify its process changes to reduce

---

[50]Breton et al., 1991, p. 391; Wever and Vorhauer, 1993, p. 21.

[51]Breton et al., 1991, p. 403; Wever and Vorhauer, 1993, p. 21.

[52]Junkins, 1994, p. 58; Williams, 1992/93, pp. 181–184.

[53]Ochsner, 1995/96, p. 73.

[54]Nagle, 1994, p. 244.

its dependence on carbonyl sulfide on cost grounds alone.[55] Similarly, Navistar could justify its process changes in its robotic paint sprayers strictly on the basis of the reduction in paint consumed in the process.[56] Alcan could justify its move from traditional furnace defluxing to in-line fluxing to remove contaminants from molten aluminum on the basis of cost savings alone.[57] And Chrysler could justify the elimination of mercury-based switches solely on the cost savings from not having to label cars containing mercury.[58]

Second, the expenditure can be justified on strategic grounds. Most often, this "strategic" explanation is simply an extended view of cost-effectiveness. If a firm had accounting systems that could capture the full costs and benefits of an environmental decision, that decision could be justified using traditional project evaluation techniques.[59] In some ways, this is more an argument for better accounting than for a more strategic vision. Other environmental activists, though, do seek a more strategic vision, one that exploits environmental issues to get a competitive advantage. Individual firms promoting this approach often want environmental concerns to be integrated with general business strategy so that they become just one more element in a broader company vision. This approach remains unusual—a recent A. D. Little survey found that only 4 percent of respondents reported that their companies "manage environmental issues as a full-fledged part of their business management approach." But 27 percent are trying to move in this direction.[60]

Whether environmental management is fully integrated with general management or not, innovative firms find another advantage in asking their people to justify environmental activities on strategic grounds. They find that it promotes the kind of personal initiative, risk-taking, and team building that can break down old mindsets.

---

[55]Ochsner, 1995/96, p. 73.

[56]Owen, 1995b, pp. 61–62.

[57]Lagoe, 1995/96, p. 53.

[58]Kainz et al., 1996, pp. 78–80.

[59]See, for example, Duke, 1994, p. 437, with citations; and Porter and van der Linde, 1995, offers many examples.

[60]Greeno et al., 1996, p. 5.

These companies see a strategic advantage to rewarding efforts to improve environmental management, even if those efforts do not generate results that would justify action in other parts of the firm.[61]

Full Cost of Environmental Decisions

Given the emphasis cost gets in justifying decisions, it is not surprising that innovative firms have given a great deal of attention to improving their cost-accounting systems to support such decisions. Many different cost-accounting systems have been developed, and confusion often occurs because people use the same names for different approaches.[62]

In large part, differences concern how many elements of cost to include and how to deal with uncertainty about these elements, especially over time. Broadly speaking, the following costs should be included:[63]

- *Usual*: Standard capital costs, operating costs, revenues from products.

- *Hidden*: Environmental costs incurred when complying with regulation, including permitting, reporting, monitoring, testing, training, operating control equipment, take-back, paying outside transport and disposal vendors, counsel, and consultants.

- *Liability*: Noncompliance penalties; future liabilities for cleanup; personal injuries; and property damage, including natural resource damage, disposal in landfills, and so on.

- *Intangible*: Effects beyond compliance, including effects on corporate image, worker morale, and market share, and concerns associated with uncertainty about the future.

---

[61]Ochsner, 1995/96, p. 74.

[62]Bailey and Soyka, 1996a, provides a useful overview of these issues.

[63]Hoffman, 1992/93, p. 2; Ferrone, 1996, pp. 108–110.

A good approach to costing should be able to estimate how these costs differ between any two alternatives. Potential "alternatives" can address[64]

- Process or product designs

- Location options for a new facility

- Materials use (e.g., raw materials, catalysts, solvents, coatings, and cleaners)

- Sources of suppliers/inputs

- Product mix and retention

- Packaging and delivery systems

- Waste management and recycling programs

- Adoption of just-in-time or build-to-order programs

- Acquisitions and divestitures

Many of these alternatives are extremely complex and subtle, and they may have broad implications. Firms will naturally want to use their standard project evaluation methods to weigh such alternatives. The evaluation system a firm uses to support environmental management should be compatible with its general system. Great economy can be achieved by focusing on differences between alternatives so that only selected cost elements are collected and evaluated.[65] Further economy can be achieved by applying an *a fortiori* principle: Once tangible costs allow a clear choice between two alternatives, there is no need to collect any further evidence that will simply strengthen the case.[66]

With these goals in mind, innovative firms seek "to expose formerly hidden costs, and tie them to specific products and components using an Activity Based Costing (ABC) methodology."[67] These firms

---

[64]Bailey and Soyka, 1996a, p. 1.

[65]Kainz et al., 1996, p. 74.

[66]Kainz et al., 1996, pp. 78–82, provides several useful illustrations of this principle.

[67]Kainz et al., 1996, pp. 73–74. ABC was originally a response to the fact that (1) traditional cost accounting allocates indirect costs in proportion to direct costs in

are now typically learning to use ABC to evaluate decisions throughout their organizations.

Even the best firms face serious barriers when they attempt to apply ABC to a specific set of questions relevant to environmental management. For example, no one has yet set up an ABC-compatible set of cost accounts that gives project evaluators easy access to data on the items listed above.[68] Innovative firms work around this problem by using the data they do have in creative ways, but the lack of data aggravates questions that swing on the importance of intangible costs and benefits, which many observers believe are more prevalent in environmental management decisions than in other decisions a firm faces.[69] Individual cases typically speak of intangibles but justify specific decisions on the basis of credible estimates of tangible costs.[70] Analyses of differences between alternatives and the *a fortiori* principle keep cost estimation simple enough to reduce the importance of these problems. In sum, proactive firms apply ABC to environmental questions by using creative analyses of specific issues.

Always present is the question of what costs and benefits to include.[71] Should a firm include benefits or costs that accrue to outside parties? Integrated supply chain analysis often identifies such flows and uses them as the basis for win-win negotiations with customers and suppliers. But if such negotiation is not available, standard project evaluations typically do not include such costs and benefits. A firm attempting to project a green profile or trying to build trust with a particular external stakeholder might take a differ-

---

product evaluations and (2) indirect costs grossly dominate direct costs in most modern industrial settings. It provided an alternative way to allocate indirect costs (Johnson and Kaplan, 1987). ABC now provides a methodology for identifying cost objects of any kind—say, production costs for an individual product; for identifying all costs associated with that object; and for allocating those costs, direct and indirect, to the object. Hence, it is well suited to linking environmentally related costs, traditionally treated as indirect costs, to specific cost objects relevant to environmental decisionmaking.

[68]Thus Marc J. Epstein's conclusion after surveying 100 leading corporations in a study conducted at Stanford's Graduate School of Business. (See Owen, 1995b, p. 65.)

[69]Ochsner et al., 1995/96, p. 68.

[70]Breton et al., 1991, p. 391.

[71]Kainz et al., 1996, pp. 74–76.

ent view. It must set the boundaries relevant to its application of cost data to be compatible with its broader strategic goals.

These subtleties emphasize that applying ABC to environmental management questions involves much more than the simple computation of a spreadsheet. As a result, only large, sophisticated firms are pursuing such cost analysis today.[72] These firms are trying many variations on the themes presented here.

Managing Suppliers

The discussion above implicitly assumes an organic activity, but all of the ideas discussed here also apply to a contract setting.

In general, integrated supply chain management seeks a close relationship between buyer and seller. It seeks a level of mutual trust that allows the parties to dissolve the boundaries between them so that they can both benefit from a more integrated view of their relationship. This thinking applies to environmental issues just as it does to other issues. The literature reports firms using four kinds of approaches to work with their suppliers more effectively.

First, buyers shorten their lists of qualified suppliers and invest more deeply in those that remain. For example, Proctor and Gamble does not produce its own packaging, but because packaging is an important part of its environmental management program, it must develop more effective partnerships with its suppliers to get the kind of environmentally friendly packaging it wants.[73] Ten years ago, Proctor and Gamble rotated suppliers as often as once every six months to maintain price discipline. It now seeks longer-term relationships— contracts run two to four years—and is working hard to develop trust in relationships that extend beyond individual contracts.

Second, firms are requiring their suppliers to demonstrate basic elements of environmental responsibility. That is, as firms reduce the number of suppliers they trade with, they are using environmental considerations to help identify those who will continue to supply

---

[72]Duke, 1994, p. 437; Nagle, 1994, p. 244.
[73]Maxwell et al., 1993, p. 328.

them. They use many different vehicles to do this, including ISO 9000, a widely used international standard that verifies compliance with basic elements of TQM and that can be extended to cite specific management processes relevant to environmental management. In Europe, firms can be registered to a variety of standards that address environmental concerns directly. They include BS7750 in the United Kingdom, and ISO 14001 and the more demanding Eco-Management and Audit Scheme (EMAS) in Western Europe as a whole. Many firms continue to maintain their own standards for the environmental performance of suppliers, presumably waiting for a satisfactory international standard to emerge.[74]

Third, firms use more complete estimates of costs to compare offers from suppliers.[75] AT&T and Ontario Hydro both apply versions of green accounting to estimate the full costs of acquiring inputs from alternative sources.[76]

Fourth, firms are inserting language in their contracts to promote proactive behavior in their suppliers like the behavior they encourage in their own firms. This language pairs objectives and metrics such as those discussed above with award fees that give suppliers incentives to respond positively to the metrics.[77]

Integration as a Basic Theme

Running through this discussion of proactive environmental management is a constant call for greater integration of environmental decisionmaking into the general decisionmaking of the firm. This point comes up repeatedly in the literature:[78]

> [A] "green wall" separates the environmental team from business functions, a wall that must be chipped away to improve environ-

---

[74]Breton et al., 1991, p. 394; Rank Xerox, 1995, p. 14.

[75]Penman and Stock, 1994, p. 842.

[76]Blumenfeld and Montrone, 1995, pp. 86–87.

[77]Marchetti et al., 1995/96, p. 47; Schmidheiny, 1992, p. 188.

[78]A 1992 Abt Associates survey of major corporations confirms this, observing that "the boundaries between environmental management and other functions have begun to blur" (Lent and Wells, 1992, p. 385).

mental—and business—success. To create companies with integrated environmental strategies, environmental managers see the need to become part of the business strategy.[79]

Environmentalism has become another component in the firm's cost-service trade-off analysis. In the same way that cost and services issues are considered when making logistics and marketing decisions, environmental issues must be examined as well. Although there may be obstacles or impediments—perceptual, technical, operation, and regulatory—to implementing . . . logistics strategies, they are not insurmountable. Maintaining focus and persistence will result in successful strategies and programs. Organizations "environmentally evolve." Firms begin by complying with environmental regulations and laws. They typically develop partial, then complete, recycling programs, which are followed by significant changes in products and packaging. Gradually, firms fully integrate environmentalism into their corporate cultures until it becomes a way of life.[80]

The AT&T/Intel benchmarking effort discussed above highlights the importance of integrating pollution prevention into product, process, and manufacturing equipment design.[81] In its review of the "best of the best," A. D. Little found that innovative companies seek to weave the environmental expectations of stakeholders into their business decisionmaking process and overall strategy. They seamlessly integrate environmental considerations into their business decisions by placing environmental expertise directly into their lines of business.[82] Environmental professionals serve as knowledge brokers to the businesses; they gather and interpret the necessary information from the external world and present it in such a way that the businesses can understand it, integrate it, and use it to take action. In this setting, environmental professionals must maintain a broad perspective, looking beyond manufacturing to the whole firm's needs.

---

[79]Kirschner, 1996, p. 19.

[80]Penman and Stock, 1994, p. 854; see also Bailey and Soyka, 1996b.

[81]Klafter et al., 1993, p. 42.

[82]Haines, 1993, p. 367; Blumenfeld and Montrone, 1995, pp. 80, 88.

Individual cases raise this point repeatedly. For example, at AT&T, "ambitious goals relating to reductions in pollutants and decreases in use of raw materials are being woven into the fabric of the business from product design to office processes to basic research to long-range business plans."[83] Chrysler is developing an environmental management system that makes environment, health, safety, and recycling a "critical part of the overall business plan of the company."[84]

Just as the notion of integration points to the presence of stakeholders with multiple interests in the firm, it also points to the importance of power within the firm and the need to build coalitions among internal interest groups to give environmental concerns appropriate weight in corporate decisionmaking.[85] Broader coalitions make it easier to see environmental concerns as compatible with core organizational concerns, thereby raising their legitimacy throughout the organization. Higher legitimacy should make environmental concerns more successful in intracorporate negotiations and should draw more effective corporate personnel to activities responsible for environmental decisionmaking.

Coalition building is easier where environmental managers can state their goals in terms relevant to others in the firm. Where a firm's customers seek green products, marketing becomes a natural ally for the environmental function. Where environmental emissions account for a significant portion of operating costs, those responsible for cutting operating costs through reengineering, quality programs, or other methods become natural allies. In practice, this means that environmental specialists must act and talk like part of the business. They must talk the talk of business,[86] and they must find ways to justify their programs and investments that are compatible with the core processes of the firm.[87]

---

[83]Breton et al., 1991, p. 389.

[84]Kainz et al., 1996, p. 71.

[85]U.S. Environmental Protection Agency, 1992, p. 15; Woods, 1993, pp. 36–37.

[86]Blumenfeld and Montrone, 1995, p. 80; Kirschner, 1996, p. 19.

[87]Ochsner et al., 1995/96, p. 68.

In practice, this also means that environmental specialists must work hand in hand with the manufacturing engineers, managers, and workers responsible for the core production processes of the firm. The environmental specialists may even be absorbed into the activities relevant to these core processes. For example, at Kodak, "responsibility is clearly assigned throughout the organization. . . . If an individual has responsibility for a process or product, that person 'owns' not only the product the unit makes, but its waste streams, its compliance-related problems, and even, in a sense, its basic design framework."[88] At Alcan, DuPont, Hewlett Packard, and Robbins, implementation of specific pollution-prevention activities has involved close cooperation among R&D engineers, environmental specialists, and workers on the line. These efforts would have failed if the players had not achieved a strong working trust in one another.[89]

An example of the options considered in a detailed pollution-prevention study helps explain why such close integration is important. An unnamed manufacturer of aircraft components weighed a variety of alternatives for reducing the effluents associated with a metal-finishing process that coated high-value components with an alloy of cobalt, chrome, nickel, aluminum, and traces of other elements. Figure 3.1 lists the basic options; those considered cost-effective are starred.

Identifying feasible options required close collaboration of engineers and workers. Environmental specialists provided information on the regulatory implications of different choices. Ranking the options required information from engineers, environmentalists, and workers about the process itself, multimedia implications, the status of external markets for waste products, technical pros and cons of treatment on and off site, uncertainty about the liabilities of off-site treatment and disposal relative to project costs, permitting costs for treatment on site, and regulatory limits on storage times.[90]

---

[88]Wever and Vorhauer, 1993, p. 27.

[89]Berube et al., 1992; Lagoe, 1995/96; Malachowski, 1991, p. 407; Ochsner et al., 1995/96.

[90]Duke, 1994, pp. 441–448, provides a detailed discussion of this case and the factors associated with it. It is worth noting that, although the analysis clearly identified cross-media effects—the possibility of moving emissions requiring control from one

---

Treat waste at the end of the pipe

1/2. Pretreat the waste stream on site to decrease liquid waste or reduce the cost of disposal (two different options)

*3. Identify industrial waste exchange to accept acid so that it is not RCRA-listed or disposed of

Improve housekeeping or maintenance

*4. Cover acid tanks on weekends, shut down hoods and scrubber to cut water consumption and air emissions

5. Use better quality assurance to reduce the number of items stripped

6. Add a second rinse tank to extend the life of rinse

Change the production process

7. Eliminate the use of 1,1,1-trichloroethane as a degreaser

8. Eliminate the use of Freon as a drying agent

*9. Reduce the volume of the acid bath

10. Use disposable masks

11. Treat and recycle baths on site

*12. Recycle scrubber water, neutralize it, and return it to the scrubber

---

Note: Asterisks denote options considered cost-effective.

Figure 3.1—Options for Reducing Effluents from a Metal-Finishing Process

What is the best way to coordinate the skills of environmental specialists, managers, engineers, and production workers in such a setting? The literature covering private-sector experiences is unclear. Specific answers vary with the circumstances, but certain patterns have emerged in recent practice. Innovative firms typically agree that[91]

---

medium to another—it did not highlight this fact or relate it to the effort under way in regulatory reform to give cross-media effects more attention. The cross-media effects were simply facts for assessment. The analysis developed estimates of costs and benefits for each option; it developed conclusions about cost-effectiveness without regard to the cross-media implications of the options.

[91]Hoffman, 1992/93, p. 5. For illustrations in the context of specific firms, see Breton et al., 1991, pp. 391–394 (AT&T); Owen, 1995a, p. 4 (Motorola, Caterpillar); and Ochsner et al., 1995/96, pp. 69–70 (DuPont). See also Klafter et al., 1993, p. 42.

- Environmental specialists should work face to face with managers, engineers, and workers responsible for core production processes. Good personal relationships among these players are important.

- The traditional owners of R&D and production associated with these production processes should retain ownership and responsibility for them. In particular, specific proposals for process changes should typically come from traditional R&D labs and engineers on site, not from environmental specialists.

- Environmental specialists should make these core owners aware of the regulatory and other environmental implications of their actions and help them become more environmentally conscious.

- Environmental specialists will be most successful in this role if they remain advocates for environmental values but learn to communicate them and justify action in language familiar to the owners of the core processes.

These considerations provide strong support for the use of cross-functional teams to integrate environmental and core business values. Firms define cross-functional teams differently. For example, DuPont organized its efforts to reduce carbonyl sulfide at its Edgemoore, Delaware, plant around a small senior team that included the unit manager for operations, the environmental manager, a senior R&D associate, and an assisting R&D engineer.[92] This team, assisted from time to time by additional players as the need arose, remained in place for at least four years as an approach was developed and refined.

When John Deere discovered hydrocarbon contamination in the ground at its Waterloo, Iowa, plant, it saw a need for a broader set of skills.[93] Table 3.2 shows the functional representation on the team it organized to develop a solution and the key concerns voiced by each function in the course of the deliberations. A detailed review of the alternatives considered and the basis for the final decision reveals that each of these functions played a role in the development of a

---

[92]Ochsner, 1995/96, p. 66.
[93]DeLange, 1994, pp. 54–55.

Table 3.2

Functional Representation on a Cross-Functional Team at John Deere

Function	Primary Functional Concerns
General management	Maintain Deere's good-neighbor image and manage life-cycle costs and reliability of product delivery.
Environmental engineering	Find a long-term solution to make sure this does not happen again; avoid registration and paperwork costs; make inspection foolproof; contain any overflows or spills.
Energy management	Keep energy use and cost down; maintain proper temperature and viscosity of fluids.
Accounting	Lower number of transactions with suppliers; allocate costs properly to users; ensure that invoiced quantities are correct.
Safety	Minimize employee danger; label products and hazards properly; separate explosive materials properly; maintain good access for emergency vehicles.
Operations	Make it easy to operate and maintain; keep labor requirements down; support reliable performance; make it easy to monitor.
Plant engineering design	Make it simple, low-energy, nonpolluting, safe, low-risk, cost-effective, and compliant with all current and foreseeable regulations.
Maintenance	Keep it simple to allow easy housekeeping.
Customers	Allow reliable shipment of fluids where needed; minimize customer's need to store bulk fluids on their sites; minimize supply management paperwork.

Source: De Lange, 1994.

satisfactory solution. Once this team had identified a solution, it was disbanded.

An important trick in cross-functional integration is finding a way to integrate the values associated with all of these functions and simultaneously preserve the organization's expertise in each of them. Different firms do this differently. Broadly speaking, however, these firms preserve career tracks in each functional specialty so that individuals can develop the knowledge required to represent their functional areas. The firms then give individuals authority to commit their functional areas to specific agreements in the contexts of specific cross-functional teams. The teams discussed above could act without continual consultation between the team members and their functional principals elsewhere in the firm.

This approach asks each functional representative on a team to balance the specialized needs of the function with the consensual needs of the group. Learning to achieve and maintain such balance creatively is a skill in its own right that benefits from experience and formal training, as discussed below.

IMPLEMENTATION

Implementation can mean two different things in the context of environmental management. The first meaning focuses on a single initiative. It considers the way an organization translates a high-level policy—for example, elimination of stratospheric ozone-depleting chemicals in production—into specific changes in processes and products that realize the central purpose of the policy. The second meaning focuses on the processes an organization uses to identify high-level initiatives, as well as the processes it uses to monitor its progress against all of its high-level goals.[94]

Both views of implementation are of interest, especially the second. Many examples have already been cited of the ways firms have executed specific environmental initiatives. The broader question about implementation raises issues that extend beyond environmental management. Any effort to change the way a large organization behaves raises a set of generic questions about implementation. The following discussion focuses on these broader questions but continues to illustrate them with examples from environmental management.

Innovative firms all agree that broad organizational change must start at the top. To succeed, a broad change must motivate and empower employees throughout the organization. This requires particular attention to incentive and training programs. To realize the kind of change discussed here, an organization must promote learning, which requires continual and intense communication throughout the organization and with key external stakeholders. Creating an environmental management program that can draw on a skilled, motivated workforce to generate and execute creative initia-

---

[94]Woods, 1993, p. 38.

tives takes time, and it benefits from a formal program that coordinates all these elements of implementation over time.

Support of the Senior Leadership

If innovative firms agree on anything, they agree that broad organizational change can succeed only with the strong and continuing involvement of the senior leadership. From this perspective, not all firms are committed to pursuing aggressive environmental programs. In 1995, Price Waterhouse found that only 40 percent of the firms responding to a survey placed oversight of environmental policy at the board level.[95] But that was twice the number that placed such responsibility at the board level only three years earlier, in 1992. Senior-level acceptance of aggressive environmental policy is spreading. Today, in more than half of the firms responding to the survey, the senior corporate environmental manager reports to the chief executive officer or another senior executive. Corporate managers take the position of the senior environmental official in the firm as a strong symbolic message about how close environmental concerns are to the core interests of the firm.[96]

As noted above, the senior leadership develops broad goals and typically leaves development of specific objectives to the major business units.[97] But the senior leadership can play a key role, even for specific initiatives. When AT&T launched its proactive corporate environmental management initiative, Chairman Robert E. Allen announced and explained it at the annual shareholder meeting.[98] This announcement was designed explicitly to deliver the message that this initiative was one of the most important priorities of the core leadership of the corporation.

Support of the senior leadership remains important even after a program has been launched. Innovative firms are setting up systems to keep their senior executives regularly informed about the environmental status of the firm. This way, the senior executives are ac-

---

[95]"The Green Machine," 1995, p. 17.

[96]Schmidheiny, 1992, pp. 192, 197.

[97]Maxwell et al., 1993, p. 322; Ochsner, 1995/96, pp. 66, 73; Owen, 1995b, p. 59.

[98]Breton et al., 1991, p. 390.

countable for environmental performance and can raise concerns if the program's implementation is not consistent with the firm's goals.

Empowering and Motivating Employees

The first step toward engaging employees in an effort to change a large organization is to make sure they know what is expected of them. That is, "responsibility [must be] clearly assigned throughout the organization."[99] The organization must then ensure that these employees have the attitudes, skills, and capabilities required to make the change work.[100] Navistar's list of the attributes that it seeks to find or instill in the employees who will implement large organizational changes are representative of the attributes mentioned by other firms:[101]

- Personal conviction

- Tenacity

- Focus and acceptance of responsibility

- Willingness to take risks

- Ability to demonstrate and reward pollution-prevention success

- "Bias for action"

Five types of formal training are important to improvements in environmental management:

1. Firms trying to raise the perceived relative importance of environmental concerns provide training about the general social importance of environmental issues and the role their firm can play in this broader setting. Such training often integrates factual presentations, emotional appeals, and open-discussion groups to try to change the attitudes or even the values of the firm's employees.[102]

---

[99]Wever and Vorhauer, 1993, p. 27.

[100]Klafter et al., 1993, p. 42; see also Richards and Frosch, 1994, p. 3.

[101]Marchetti et al., 1995/96, pp. 49–50

[102]Penman and Stock, 1994, p. 842; see also Richards and Frosch, 1994, p. 23.

2. Firms using new management methods (such as cross-functional teams) to promote integration train their employees in the use of these teams and in more general consensus building and problem-solving techniques relevant to their success.[103] Similar training is important to any manager being asked to be more creative and persistent about environmental issues, although general management experience is often the best teacher of these skills.

3. Firms seeking to develop environmental specialists who can operate confidently in many aspects of environmental decisionmaking—for example, specialists who can function effectively as decisionmakers on cross-functional teams—develop databases that these employees can use for self-paced instruction. Such databases offer current information on technologies or case studies of past decisions that young employees can access as needed when they face specific problems in their day-to-day work.

4. Firms that are facing new regulations, introducing new pollution-prevention programs, or adopting new databases or analytic tools offer targeted training to the employees that these changes will affect most directly.[104]

5. Firms seeking to establish a critical mass of expertise on environmental issues that can sustain their experts over time and help them work together to keep their skills up to date form centers of excellence or competence centers. Such centers can support the training options listed above and provide points of focus for longer-term career development.

Training takes time. The more interaction the firm seeks between trainer and trainee, the more time-intensive is formal training. Informal on-the-job training related to the execution of new programs and procedures continues indefinitely.

Every firm seeking to improve its environmental management gives special attention to incentives. Firms tend to choose incentives that are compatible with their prevailing corporate cultures. Depending

---

[103]Woods, 1993, p. 35.

[104]For a good example, see Megna and Savoy, 1991.

on the culture, incentives target individuals, teams, or organizations; they can be direct or indirect, monetary or nonmonetary.[105]

The most common form of incentive mentioned in the literature is a direct, nonmonetary award to or recognition of individuals who have tangibly improved environmental management. Greater achievements are recognized at higher levels in the firm.[106] Firms emphasize the importance of giving such awards frequently, even for small improvements, to diffuse the importance of environmental management throughout the organization.[107] Firms that offer bonuses for large improvements include environmental management in such programs.[108] Other firms refuse to pay individual employees cash for "doing their job." For them, "formal awards are less important . . . than corporate and plant norms that reward personal initiative, networking, and team building."[109]

Many firms point to the importance of placing key environmental management positions on a promotion path that attracts highly qualified managers and rewards good performance with promotions.[110] Team-oriented firms can use formulas written to reflect environmental management activities to allocate profit-sharing bonuses to team members. Some companies argue less directly that cost-effective environmental management improves overall corporate profits and that all employees benefit through profit-sharing arrangements.

Some firms try to "internalize" relevant costs by allocating compliance and other environmentally related costs to business units before assessing unit profits. These units are then encouraged to allocate such costs to product lines for pricing and product-evaluation purposes.[111] Others tax business units for using more than a certain number of external sources for disposal or recycling services. Rev-

---

[105]Hoffman, 1992–93; Ochsner, 1995/96, pp. 70–74; Wever and Vorhauer, 1993, p. 21.

[106]Klafter et al., 1993, p. 42.

[107]Woods, 1993, p. 36.

[108]"The Green Machine," 1995, pp. 17–18.

[109]Ochsner et al., 1995/96, p. 70.

[110]Ochsner et al., 1995/96, p. 67.

[111]Breton et al., 1991, p. 391; Owen, 1995b, pp. 59, 65.

enues collected from such internal taxes can be used to recover environmental costs formerly allocated to overhead or to generate investment funds that can be committed to pollution prevention and other proactive programs where they are collected.[112]

In all of these approaches, firms typically use competition among business units to heighten the incentive effects of options and to gather internal benchmarks that can be used to allocate incentives. In sum, all proactive firms recognize the importance of incentives to successful implementation of environmental management, and each firm uses the specific incentives it is most comfortable with in its cultural setting.

A special incentive issue that proactive firms recognize is the challenge of dealing with failed pollution-prevention experiments. Trial and error offers great potential in any learning organization, and it is especially important in efforts to refine changes in an ongoing production process. Systematic learning depends on a system that supports flexibility and tolerates the right kinds of mistakes.[113] The most important aspect of successful experimentation is to recognize that failure is part of the learning process. The term "failing forward"— that is, "creating forward momentum with the learning derived from failures"—usefully describes this process. Operationalizing this concept requires distinguishing between intelligent failure and unnecessary failure and setting up systems to learn from both.[114]

Communicating Continuously in All Directions

Proactive firms agree that continuous communication, in all directions, about the goals and status of the environmental management program is important to success.[115] Such communication serves four goals:

---

[112]Marchetti et al., 1995/96, p. 47.

[113]Ochsner et al., 1995/96, p. 71.

[114]Leonard-Barton, 1996, p. 119.

[115]Junkins, 1994, p. 58; Richards and Frosch, 1994, pp. 2, 23.

1. It conveys to the whole organization the senior leadership's commitment to effective environmental management. This occurs first when the firm moves toward a new commitment to environmental management and then repeatedly over time to verify continuing support.

2. It conveys knowledge about the realized performance of environmental management to the senior leadership so that they remain accountable for the environmental performance of the firm as a whole and can make adjustments as needed to ensure that the program being implemented continues to reflect corporatewide goals. Not incidentally, such communication maintains the awareness of those in senior management, contributing to their continuing willingness to support environmental management efforts in the broader context of their responsibilities.

3. It conveys information on successes from one business unit to another in order to maintain the momentum of change and support learning across the organization.[116] It conveys information about failures, with similar objectives. Failures can threaten the program, especially early in its life, if the firm does not react to them constructively. Communication about failures is most successful when it also includes a constructive corporate response.

4. It conveys information on the goals and status of the program to key stakeholders outside the firm. Depending on its corporate vision, the firm can give special attention to customers, regulators, or local communities, including its employees who live in those communities.

The Time Required for Change

Any attempt to induce a large change in an organization takes time. First, a case for change must be built and taken to the senior leadership. Following leadership signoff, teams must be formed to articulate policy and transform it into specific, implementable actions. This often requires experimentation and prototyping to test the new policy in a way that limits the effects of failure. More often than not, more than one effort is required to create an implementable action

---

[116]Breton et al., 1991, p. 403; Klafter et al., 1993, p. 42.

that works as intended. Failure more often results from an unwillingness to accept the change than from any inherent technological flaw in the change itself. The effort to achieve implementable change may raise unexpected questions that need to be resolved at senior levels before work can proceed. Such questions often arise when the importance of units not originally involved in the change becomes apparent only after implementation starts. Their role in the change must then be brokered, often at a high enough level to encompass all of the affected offices.

These problems are not unmanageable. On the contrary, this general pattern of problems is quite predictable and can be managed in fairly routine ways in organizations that have become used to such change. These organizations often have many such changes under way simultaneously in different areas. They break change down into manageable sets and test one set before going on to the next. In a learning organization, such change never ends; it becomes a normal part of standard operating procedures.

Nonetheless, innovative organizations recognize that change takes time. For example, when Weyerhaeuser wanted to expand its management focus on cleaners, paints, and other chemical-based supplies, a fairly narrow environmental management task, it formed a task force in 1992 to articulate and document the direction for the program.[117] After a year, it had a clear enough program to take to senior management. After receiving senior management approval, Weyerhaeuser scheduled another full year for "missionary" work at the 200 locations of the firm in North America to increase awareness of the problem and to suggest simple solutions. The program actually began in the field in 1995 and achieved full implementation across the firm only in 1996.

To develop and introduce a single new green product/packaging combination, Proctor and Gamble typically finds that it needs two to

---

[117]Loren, 1996, pp. 101–106.

six years.[118] To reduce the use of carbonyl sulfide at one plant, DuPont committed a significant portion of the time of members of the responsible senior-level cross-functional team over a four-year period.[119] To reduce chlorine use at a single plant, Alcan worked with equipment manufacturers and others for over six years, refining target processes on site.[120] Table 3.3 lists a series of milestones in the Alcan effort, reemphasizing the importance of integrating environmental with core production concerns. The Texas Instruments Defense Systems and Electronics Group took three years of focused effort, with extensive feedback from outside auditors, to implement a quality-based resource management system that embodied elements like those discussed above.[121]

Specific changes affecting small parts of an organization—say, a few hundred people—may take as much as two years to work through. Implementing a specific change throughout an organization with tens of thousands of employees can easily take five years or more. Implementation of a new approach to environmental management typically involves a series of specific changes that can extend the period of change beyond a decade. And again, in the limit, change in a learning organization never ends.

This pattern of change has two important implications for the DoD. First, all commercial firms are small relative to the DoD. Efforts to induce change throughout the DoD will in all likelihood take even longer than they do in any large commercial firm. Second, the senior leadership in the DoD and the armed services typically has shorter tenure than the leadership in large commercial firms. It is very unlikely that any leadership team in the DoD could see a significant organizationwide change through from conception to full implementation during its tenure.   Hence, to achieve significant organizationwide change, the DoD must give even more attention to institutionalizing the change *process* than commercial firms do.

---

[118]Maxwell et al., 1993, p. 323.

[119]Ochsner et al, 1995/96, p. 66.

[120]Lagoe, 1995/96, pp. 54–55.

[121]Junkins, 1994, p. 57.

Table 3.3

Pollution-Prevention Activities at the Alcan Oswego Plant

Description of Change	Date	Reduction from 1986 Emission Levels (percent)
Installed Apur in-line fluxing unit on holding furnace no. 5.	May 1987	9
Reduced chlorine in fluxing gas from 50% to 30% for can body stock alloys.	December 1987	15
Eliminated fluxing in melting furnaces 1, 2, 3, 4, and 5.	June 1988	29
Reduced fluxing time in holding furnace no. 5 from 30 to 10 minutes per cycle.	March 1989	2.5
Installed Union Carbide R-180 in-line fluxing unit on holding furnace no. 4.	July 1990	8
Reduced chlorine in fluxing gas from 30% to 20% on holding furnaces nos. 4 and 5.	July 1990	3
Reduced fluxing time on holding furnace no. 4 from 30 to 10 minutes per cycle.	July 1991	2.5
Commissioned melting and holding furnace no. 6 and removed melting and holding furnaces nos. 1 and 2 from service.	September 1991	14
Reduced chlorine in fluxing gas from 30% to 20% on holding furnace no. 3.	August 1992	2
Reduced chlorine flow rate from 1.2 to 0.9 pounds/minute on holding furnace no. 6.	October 1992	1
Implemented furnace operator monitoring of chlorine usage.	January 1993	8
Computerized chlorine consumption monitoring on melting and casting centers nos. 4, 5, and 6.	October 1993	1

A Formal Program for Environmental Management

A full-fledged environmental management effort could easily have many objectives in place at one time, with multiple efforts throughout the organization addressing each objective. As objectives evolve over time, these efforts change as well. Given the periods of time required to execute any single effort, proactive firms recognize the value of placing environmental management in a formal program.

Programs differ from firm to firm, but all of them point to an advocate or champion for environmental management on the corporate staff and on the staff of each business unit and plant.[122] These individuals and their dedicated staffs have the resources to oversee environmental management in a firm and ensure its success. They serve as the environmental link between the senior leadership and the core production activities of the firm. They convey the corporate goals to these activities, help the owners of these activities develop metrics that reflect their environmental performance relevant to the corporate goals, use these metrics to monitor the performance of the core activities, and report performance to the senior leadership on a regular basis.

They also work with the owners of core activities to develop longer-term environmental plans to help the firm make decisions about the use of its investment and analytic resources over time. Such plans provide a useful way to assist decisionmakers in thinking through the sequence of incremental steps needed to achieve long-run environmental management goals. For example,

- Proactive firms choose specific changes that they can build on for future learning. Firms that own multiple plants and pursue multiple activities at each plant may target environmental management of one activity in each plant.[123] One activity serves as a seed for extending change to analogous activities at other plants, as well as to other activities at the same plant.

- To identify such seed activities at plants, some firms seek lower-risk, higher-payoff changes first.[124] This approach uses early successes to give change momentum.

- Proactive firms develop a well-defined plan for each change at the location where the change will occur. The plan may extend over several years and may change during that period, but it changes against a set of planning milestones that provide a basis for accountability.

---

[122]Klafter et al., 1993, p. 42; Ochsner et al., 1995/96, p. 69.

[123]Rank Xerox, 1995, pp. 22–23.

[124]Owen, 1995b, p. 59; Bailey and Soyka, 1996b, pp. 16–19.

No specific change locks the organization into a future it cannot alter. Successful firms recognize the need to continue looking beyond each specific change to a broader commitment to flexibility and learning. Proactive commercial firms often speak of effective environmental management as being more an attitude and an approach than a formal program. In these firms, effective environmental management becomes a commitment to seek high-quality performance and then sustain it through continuing learning and improvement.

McKinsey and Co. developed a simple approach to implementing innovations in logistics activities that should be useful to environmental management in the DoD as well. This approach proceeds in three steps:[125]

1. Develop an operational blueprint. Identify an overall goal, the operational changes required to get there, and needed changes in systems and procedures. This involves an intensive examination of current operations by line managers and staffs to identify and test alternatives.

2. Assess organizational readiness in terms of structure, skills and tools, and information systems. Check links between sales and manufacturing, as well as the general information flow. Start with top-down direction, but work out the details with bottom-up initiatives.

3. Implement by having middle managers identify improvement goals and execute action plans to achieve them. Success depends on having everyone buy in. Performance metrics at the department level and one level up, reflected in performance reviews, are key to success. Create a formal process of self-assessment to link teams to their self-stated goals and track performance. Only then turn to detailed design work, with user involvement.

A successful firm applies such a systematic approach to each change it makes. Environmental management program leaders must ensure that every change in their program follows such a systematic path.

---

[125]Rose and Sharman, 1989.

The DoD has a great deal of experience with formal programs of change management. Modifications induced by continuing learning occur throughout the life of any major weapon system. At any point in time, under normal circumstances, the System Program Office (SPO) may be managing tens or hundreds of modifications. "Basket" SPOs that manage a family of related systems, such as engines or transport aircraft, remain in place indefinitely, managing the development and modification of one system after another.[126]   The champion for change actions associated with environmental management might use a similar structure to manage the many elements of such changes. Environmental management is about organizational as well as technological change, since organizational acceptance of a specific change is at least as important to success as technical response to a combat contingency is in a traditional weapon SPO.

TOTAL QUALITY MANAGEMENT AS A SOLUTION

Anyone familiar with TQM will have noted by now how much environmental management in innovative firms sounds like TQM. As noted in Chapter One, TQM seeks to (1) identify an organization's customers and what they want, (2) identify the processes in the organization that serve the customers and eliminate as much waste— activity that does not add value for the customer—as possible from those processes, and (3) monitor performance against the first two goals and continually improve that performance over time.[127] "Quality" refers to anything the customer values. TQM seeks to eliminate any activity that does not contribute to quality, so defined.

If we substitute "stakeholders" for "customers," much of what innovative firms are doing is completely consistent with this description of TQM. These firms recognize this,[128] and indeed, individual cases reviewed by RAND contain repeated references to TQM. Many of the

---

[126]Camm, 1993, describes how such an organization works. See also Leonard-Barton, 1996, p. 92, on managing the implementation of organizational change as an innovation process.

[127]Levine and Luck, 1994.

[128]For a useful collection of quality-based perspectives on environmental management, see Willig, 1994.

firms in these case studies think of themselves as "quality organizations" that are simply applying their broader corporate TQM culture in the area of environmental management. Many of the firms talk about using *formal* TQM templates to implement a quality approach to environmental management. Even firms that do not speak explicitly about quality have in many cases been using TQM concepts for many years. It is quite possible that they no longer talk directly about TQM because it is now so embedded in their processes that they no longer distinguish it from their normal way of doing business. The following discussion briefly notes several specific commercial experiences with TQM and then describes the four formal quality templates most relevant to innovative environmental management in the United States.

TQM as a Normal Approach to Business

A 1992 Abt Associates survey of major nonservice corporations found that

> TQM technique is being rapidly adopted by environmental management, which is providing environmental efforts with a path to wider legitimacy and top management acceptance. Because it follows in TQM's footsteps, pollution prevention is a natural. Like TQM, pollution prevention gets its start on the plant floor by looking at process improvement and design. Both efforts share a hatred for inefficiency and a focus on a continually improving ideal: in TQM's case, zero defects and in pollution prevention's, zero pollution. Ultimately, both fields aspire to the same ideal, zero waste in delivering value to the customer. . . .
>
> Environmental management seems to be moving deeper into the corporation via corporate TQM programs, and TQM programs are being increasingly used to manage corporate environmental efforts. The two fields are cross-pollinating and creating a shared language across the firm. Sixty percent of the firms in our sample had either a corporate TQM program that was beginning to adopt environmental measures or an environmental program that was adopting TQM. . . . TQM is providing environmental management with a language of performance and quality that is widely understood by top management. . . . TQM is bringing to environmental management a customer focus that sets the stage to tie environmental manage-

ment into firm performance on the revenue side of the equation. If this trend continues, environment can be managed as a profit center.[129]

Rank Xerox is a good example of a quality organization that has adapted its general approach to quality to environmental management: "The company's environmental programme is an integral part of its TQM process which emphasises continuous improvement and teamwork." The "total quality culture that has been established within Rank Xerox for more than a decade has resulted in dramatic improvements in all aspects of the business, particularly in waste reduction." Its "Environment, Health, and Safety Management Model uses the same business quality structures, processes, and principles as the Leadership through Quality strategy and is consistent with the requirements of external environmental management registration systems such as BS7750, EMAS, and the proposed ISO 14000 series."[130] Its method of documenting a "waste-free factory" draws heavily on the Baldrige Award approach (see below).

Similarly, AT&T has aligned its environmental objectives "with implementation of the TQM process within the company. This approach is consistent with the philosophy that an environmental incident is an indication of a process failure. Waste and emissions are to be treated as defects in the quality process. Through the Quality Policy Deployment Process, the environmental [objectives] are being elevated to a level comparable to that of product performance, reliability, and price."[131]

Mobilizing to eliminate CFCs in manufacturing at Motorola, says Robert C. Pfahl, Jr., director of advanced manufacturing technology at Motorola, "was like introducing TQM. Companies with established continuous-improvement, TQM, or ISO 9000 programs know

---

[129]Lent and Wells, 1992, pp. 390–391. See also U.S. Environmental Protection Agency, 1996b, pp. G6–G7.

[130]Rank Xerox, 1995, pp. 6–7, 25. BS7750 is the British standard based on ISO 9000 for auditing the environmental management systems of individual organizations. The Eco-Management and Audit Scheme (EMAS) is a similar standard developed for use in the European Community. As noted above, ISO 14000 is an international variation on these European systems. (See, for example, Jackson, 1995, pp. 66–67.)

[131]Breton et al., 1991, p. 391; Thompson and Rauck, 1993, p. 374.

what Pfahl means. The ground rules are the same: make no assumptions, measure and evaluate everything, find the costs, and wring them out of the operation."[132]

Similarly, Hyde Tools, Inc., of Southbridge, Massachusetts, used TQM to seek a "zero emission solution." This company, which manufactures knives and industrial saws, had a TQM program in place and used it to walk backward through all its environmentally relevant processes, moving from the effluents to the causes of the effluents in each process.[133]

Dow Chemical uses TQM on an ongoing basis. It has "explicitly identified the link between quality improvement and environmental performance by using statistical-process control to reduce the variance in processes and to lower waste."[134]

At Weyerhaeuser, "developing the chemical management plan vision and criteria entailed a liberal amount of common sense. Like ISO 9000 or TQM, the concepts themselves are not complex. What is difficult is getting everyone in a diverse and far-flung organization to join together toward a common purpose within a given time frame— and organize a program that will actually work."[135]

Perhaps the biggest problem firms have had with using TQM is its similarity to common sense. What exactly *is* TQM? Each firm quoted here uses a version of TQM that addresses its particular needs. But many other firms that have attempted to use TQM concepts have met with little success. A system of ideas so close to common sense has often proven elusive, which is why so many firms eventually come to appreciate the power of formal TQM templates.

Formal TQM Templates

Formal TQM templates lay out explicitly what a firm must do to guarantee that it is in fact using TQM. These templates depend

---

[132]Owen, 1995a, p. 4.

[133]Owen, 1995b, p. 63.

[134]Porter and van der Linde, 1995, p. 125.

[135]Loren, 1996, p. 102.

heavily on the use of audits to examine a firm's management processes. If the processes are properly designed and executed, TQM falls into place. The audits involved can be internal or external. Many firms find that to guarantee that they are designing and executing their management processes properly, they must accept the assistance or the burden (depending on your point of view) of external auditors trained and certified to recognize appropriate management processes.

For environmental management in the United States, four formal templates are relevant:[136]

- The Malcolm Baldrige Award

- The ISO 9000 series

- The Total Quality Environmental Management (TQEM) matrix

- The ISO 14000 series

The *Malcolm Baldrige Award* was established by Congress in 1987 and has since been administered by the U.S. Department of Commerce to recognize American organizations that exemplify the very highest level of achievement with regard to TQM. The Baldrige Award provides a formal way to judge an organization's performance against seven weighted criteria:[137]

- Leadership

- Information and analysis

- Strategic planning

- Process management

- Human resource development

- Results

- Customer/stakeholder satisfaction

---

[136]Other standards comparable in scope and intent to ISO 14000 include BS7750 in the United Kingdom and EMAS in western Europe.

[137]Wever, 1996b, p. 31.

The award was a reaction to general assent that Japan's analogous Deming Award had been instrumental to the diffusion and refinement of TQM in that country; Congress hoped that the Baldrige Award would encourage similar progress in U.S. organizations.

The Baldrige Award is highly regarded. J. M. Juran, one of the leading American champions of TQM, has said that "Total Quality Management consists of those actions needed to get to world-class quality. Right now, the most complete list of those actions is contained in the criteria for the Baldrige Award."[138] Texas Instruments' Defense Systems and Electronics Group found that the "real benefit of applying for the Baldrige Award lies in adopting its quality criteria." The Baldrige Award forced the group to think systematically about its customer, which made the customer the centerpiece of daily activities, led to better communication with customers and employees, and introduced concepts of benchmarking and stretch goals that "get an organization to world-class levels quickly and not by 5 percent to 10 percent improvements a year each year."[139]

Where the Baldrige Award seeks to identify the very best organizations in America, the *ISO 9000 series* seeks to register organizations worldwide that use TQM well enough to be reliable suppliers to other firms.[140] The International Standards Organization introduced its first management standards, the ISO 9000 series, in 1987. Different elements of the series address different kinds of organizations and various aspects of auditing performance. Audits seek to verify that an organization has operative management processes in place to address the kinds of factors identified in the Baldrige Award criteria.

The ISO 9000 standards have met with great success since their introduction. Large industrial firms increasingly use ISO 9000 as an integral part of their qualification tests for suppliers. ISO registrations are now prominently featured in the advertising in trade magazines. In *Quality Progress*, a leading trade magazine on TQM itself, the number of firms listing ISO 9000 registration rose from 23 percent in

---

[138]Juran, 1994, p. 64.

[139]Junkins, 1994, p. 57.

[140]Many guides are available. See, for example, Johnson, 1993.

1992 to 41 percent in 1994.[141] Even the U.S. automobile industry, which has maintained detailed qualification standards for decades, recently adopted a variation on the ISO series called QS-9000.[142]

Total Quality Environmental Management (TQEM), broadly understood, is simply the application of TQM to environmental concerns. The Japanese have been using TQEM since the early 1970s.[143]

The first systematic efforts to apply TQM to environmental issues in the United States began in 1990 on two fronts. More than 20 large, proactive firms banded together at that time to form the Global Environmental Management Initiative (GEMI), an organization committed to the application of TQM to environmental management in a way that will "help business achieve environmental, health, and safety excellence."[144] GEMI holds conferences and publishes basic manuals to help diffuse innovative thinking about TQEM.

Meanwhile, the Council of Great Lakes Industries began a project in 1990 to create a clearly structured approach to TQEM. Building on a TQM self-assessment matrix approach developed by Eastman Kodak (also a member of GEMI), this project ultimately yielded the TQEM matrix and accompanying self-assessment questions in 1993.[145] The matrix and the questions provide a systematic way for an organization to apply the Baldrige Award approach and criteria to its environmentally relevant processes. The matrix is explicitly designed for self-assessment. But if systematically applied, it leads an organization to compare itself with the best in the world for each Baldrige criterion and hence provides a systematic vehicle for monitoring continuous improvement toward world-class performance.

Just as the TQEM matrix takes the Baldrige Award as its lead, the *ISO 14000 series* follows the ISO 9000 series. The International Standards Organization published ISO 14001 in 1996 and published a series of additional elements of ISO 14000 over the next few years. Unlike ISO

---

[141]Stratton, 1994, p. 5.

[142]Perry Johnson, Inc., 1995.

[143]Wever and Vorhauer, 1993, p. 19.

[144]Global Environmental Management Initiative, 1992.

[145]Wever, 1996b, pp. 207–246.

9000, ISO 14001 does not require the use of external auditors; it is viewed as a guideline that any organization can use to verify that it has effectively applied TQM methods to its environmentally related processes. A number of firms, however, are exploring the ISO 14000 series as a basis for external registration. Such registration could identify them as green firms for customers seeking green products. It could also identify them to regulators seeking to encourage a more proactive approach to environmental management. Several regulatory agencies are exploring the possibility of applying regulations to firms registered to the ISO 14000 standard that are different from the regulations those firms face today.[146]

All four of these quality templates are closely related.[147] Very generally speaking:

- The Baldrige Award and ISO 9000 look at all management processes, while the TQEM matrix and ISO 14000 focus on environmental management processes.

- ISO 9000 and 14000 set minimal standards for implementing TQM, while the Baldrige Award approach and TQEM matrix are explicitly designed to allow firms to trace continuous improvement from sub-ISO performance to world-class performance.

- Users of these templates find that it is easiest to contemplate ISO 14000 if they are already registered to ISO 9000. One is a natural extension of the other. A similar relationship is likely between the Baldrige Award approach and the TQEM matrix.

- The ultimate intent of TQM is to integrate environmental management into the core management functions of an organization. As that occurs, distinctions between ISO 9000 and 14000 and between the Baldrige Award approach and the TQEM matrix should fade away.[148] No one has reached that point yet.

---

[146]As noted above, considerable controversy surrounds the use of ISO 14000 for self-assessment or supplier registration, or in a regulatory setting. See the discussion and references at footnote 21 in Chapter Two.

[147]For a detailed discussion and comparison of all four, see Wever, 1996b, pp. 17–33.

[148]See Owen, 1995b, p. 64.

- These templates are not ends in themselves. As the Texas Instruments example above emphasizes, they are all means to an end—verifying that an organization has in fact implemented TQM principles appropriately.

In the end, each of these templates is extremely demanding. Applying any one of them consumes substantial resources, and they may take years to put in place in an organization. Even firms committed to TQM often question the net value of these templates, given the costs associated with them. That said, successful and innovative firms are giving the templates close attention today.

A summary view might be that ISO 9000 offers a useful entry point to TQM for any organization.[149] It is worthwhile for any firm that has sought to implement TQM on its own and failed. Once a firm is comfortable with ISO 9000, ISO 14000 then begins to look like a natural extension. The Baldrige Award becomes more interesting, because it helps the organization achieve levels of performance well above those contemplated in the ISO standards. Firms applying the Baldrige Award approach can look back on the ISO standards and question their usefulness. The TQEM matrix offers similar promise in the specific arena of environmental management. The very best firms will migrate toward the Baldrige Award approach, and as they integrate environmental and core concerns, they will find the TQEM matrix one natural element in a broader Baldrige Award approach.

The TQEM Matrix as a Reflection of a Continually Improving Consensus

Earlier in this chapter, three variations on the general consensus developing about proactive environmental management were discussed. The TQEM matrix provides a useful addition to this list. It formally recognizes TQM as the source of much thinking about proactive environmental management today. Formally based on the Baldrige Award approach, the the TQEM matrix provides insights that are fully compatible with the general consensus.

---

[149]Owen, 1995b, p. 64.

The basic logic behind the TQEM matrix can be summarized as follows:[150]

- Know your stakeholders.

- Get senior leadership on board.

- Gather data and analyze them; funnel information into strategic planning.

- Set goals and objectives, with metrics.

- Using training and motivation; empower the workforce to work through teamwork and consensus.

- Set up quality assurance systems as feedback loops to drive continuous improvement.

- Achieve environmental results that meet stakeholder needs.

Taken together, these points provide a succinct summary of the discussion in this chapter. The next chapter explores in greater detail how ISO 14001 addresses this logic.

---

[150]Wever and Vorhauer, 1993, p. 21.

Chapter Four

USING ISO 14001 TO ENHANCE AN ENVIRONMENTAL MANAGEMENT SYSTEM

Following the pattern set by the ISO 9000 standards, ISO 14001 got off to a quick start after it was introduced in 1996. By the end of 1997, about 2,500 facilities had been registered to the standard. Eighty-five of those, including Digital Equipment, Ford, IBM, Lockheed Martin, Lucent Technologies, 3M, and United Technologies, were in the United States.[1] By June 1999, 10,700 firms had registered worldwide. Of these, 480 were in the United States. The most frequent registrations have been in the electronics industries (high-tech computers, 19 percent; consumer electronics, 18 percent), defense/aerospace (12 percent), communications (5 percent), chemicals (15 percent), utilities and power (8 percent), and the automotive industry (5 percent).[2]

Table 4.1 lists the subsections of ISO 14001 that are relevant to the development or review of an environmental management system (EMS).[3] These standards provide a structured way to define a clear corporate environmental policy and then translate it into a reliable system to ensure ongoing compliance, detect failures quickly and correct them, and, if desired, identify and exploit other opportunities that go beyond compliance and that can help an organization pursue its core interests in areas relevant to its environmental policy.

---

[1]Cascio and Hale, 1998; Woodside et al., 1998, p. ix.

[2]Reinhard Peglau, reported to ISO World at www.iso14000.com on August 26, 1999.

[3]For details on these subsections, see Jackson, 1997, or Woodside et al., 1998.

Table 4.1

Environmental Management System Requirements in ISO 14001

Subsection	Topic
4.1	General requirements
4.2	Environmental policy
4.3.1	Environmental aspects
4.3.2	Legal and other requirements
4.3.3	Objectives and targets
4.3.4	Environmental management program
4.4.1	Structure and responsibility
4.4.2	Training awareness and competence
4.4.3	Communication
4.4.4	Environmental management system documentation
4.4.5	Document control
4.4.6	Operational control
4.4.7	Emergency preparedness and response
4.5.1	Monitoring and measurement
4.5.2	Nonconformance and corrective and preventive action
4.5.3	Records
4.5.4	Environmental management system audit
4.6	Management review

As noted in Chapter Three, different organizations continue to explore and apply ISO 14001 for different reasons. For regulators, it offers a base-level standard they can build on to register willing firms as part of their design of new regulatory systems with multiple tiers; firms registered to the standard have greater freedom to seek a performance-oriented approach to compliance with the laws in a jurisdiction. For multinational corporations, ISO 14001 is a common standard that can provide a global consensus on environmental management. This could be useful for regulatory purposes or for registered organizations to certify to customers that they are environmentally responsible. Still others see it as a template they can use to improve their existing EMSs in ways that give them greater visibility, greater control over compliance, or more proactive environmental practices.

These different motives lead to different approaches to ISO 14001. Those seeking formal certification of some kind have a third-party auditor register them to the standard and continue to audit their operations to sustain the registration over time. Those seeking only a template for improving their EMSs see such formal registration as a costly exercise with few incremental benefits. Organizations seeking interim positions may self-declare their compliance with the standard but not use an external auditor to verify their initial and continuing compliance. Or they may prepare the ground for future formal registration, with a third-party auditor, if the desirability of formal certification increases in the future. Formal registration, using a third-party auditor, has been more common outside the United States than inside. Firms in the United States that pursue this option typically have extensive activities in Europe or Asia, where regulators and customers have given more aggressive support to the standard. (The DoD is in similar circumstances.) Most major private firms in the United States have chosen not to pursue formal registration with a third-party auditor.[4] Many of them reserve the right to move to full, formal registration if circumstances warrant such a change, and they have developed an infrastructure to do this quickly if necessary.

This chapter reviews the benefits that commercial firms have associated with using ISO 14001 to enhance their EMSs. It uses a simple internal audit of the Air Force EMS to illustrate that ISO 14001 could potentially provide the DoD with many of the benefits perceived by industrial firms in the private sector. Information about this audit also provides additional insight into the factors that ISO 14001 emphasizes in the design of an EMS. The chapter then reviews the experience firms have had to date implementing efforts to register some or all of their industrial activities to the ISO 14001 standard.[5] It

---

[4]For a useful compendium of explanations, see Krut and Drummond, 1997.

[5]The material presented here draws on corporate environmental management information from Eastman Kodak, Ford, Hewlett Packard, IBM, Lockheed Martin, Lucent Technologies, Rockwell Automation, Texas Instruments, and Xerox. These are all large American firms with (1) industrial processes comparable to those in DoD central logistics facilities and (2) a strong interest in ISO 14001. The section integrates information from Advanced Waste Management Systems, Inc., Bureau Veritas Quality International, MGMT Alliances, SGS International Certification Services, and Transformation Strategies, all firms that provide consulting, auditing, and/or registration services on ISO 14001. And it uses material from the following documents: Butner, 1996a; Cascio and Hale, 1998; Cascio et al., 1996; Diamond, 1996; Jackson, 1997a,b; Krut Carol Drummond, 1997; Minner, 1997; Sheldon, 1997; Wilson, 1998; and Woodside et al., 1998.

describes the different approaches that two environmentally pro-active American firms, Ford and IBM, have taken to achieve global registration to ISO 14001.

POTENTIAL BENEFITS THAT COMMERCIAL FIRMS PERCEIVE IN ISO 14001

All large commercial industrial firms reviewed ISO 14001 while it was being developed and after it was introduced in 1996. Many of the best firms found that their EMSs already exceeded the requirements for the ISO 14001 in many areas, but even when that was true, most found at least some areas where the standard helped them improve. Where those areas are and how important they are depend in part on each firm's priorities and how it intends to use ISO 14001. Some will use the standard as a template for assessing and improving their own EMSs, while others intend to be registered. Some will use a third-party auditor for registration; others will self-declare. Some will do these things at all their facilities; others, only at selected locations.

Commercial industrial firms have different perceptions about ISO 14001. These perceptions help to explain subtle differences among reasons for decisions regarding the standard.

- ISO 14001 offers a general template that any firm can use to as-sess the adequacy of its planning and management of environ-mental issues, whether the firm intends to register to the stan-dard or not. This is a very broad benefit. The template identifies the nature of a firm's environmental exposure and forces the firm to look ahead and plan for changes. This in itself can help a firm reduce its exposure, which is a broadly held concern.

- ISO 14001 offers a formal way to improve an EMS adopted for any reason by highlighting key elements of the system and track-ing progress on them. Different firms have emphasized knowl-edge of exposure to an environmental risk; knowledge of relevant laws and regulations and changes in them; formal goals, objec-tives, and targets; measurement and monitoring activities; training and awareness; auditing skills; document control; and management review.

- ISO 14001 offers a formal system that improves compliance and reduces the management resources required to comply, especially in the face of continually expanding and changing laws and regulations. Documentation helps sustain compliance as staff turns over and helps diffuse knowledge about changes as they occur. Cost-effective management of data maintains management focus. Effective management of suppliers simplifies internal management.

- ISO 14001 could in the future support a different relationship with regulators that reduces joint compliance costs and thereby improves overall environmental performance. Some firms are waiting for evidence that this will occur before registering but want to be able to move quickly if and when it appears.

- ISO 14001 offers a formal way to identify and support environment-related investments that improve businesswide performance on straight business criteria—adjusted to ensure that all costs and benefits are reflected in the assessment. Various firms emphasize pollution prevention, waste minimization, design for the environment, or supplier management (in this last case, the "investment" is in a long-term relationship). Costs include emission fees, waste disposal, control devices, cost of risk in capital associated with exposure, and lost market share.

- ISO 14001 offers a formal way to certify environmental responsibility to stakeholders who care about environmental performance. This is increasingly important as expectations change without warning. Firms emphasize better access to customers (including attraction, retention, better working relationships, and effective responses to competitors' actions), especially international customers, but also U.S. customers who want to work with their sources on environmental issues. Many firms delay actual registration, waiting for these demands to materialize, but they want to be able to move quickly when they do

- ISO 14001 offers a formal way to track progress or continuous improvement that supports sustainable development, design for the environment or green products, pollution prevention, waste minimization, or any other movement beyond compliance, for whatever reason.

- ISO 14001 offers a formal way to integrate environmental concerns with broader business concerns. In particular, it makes all parts of a firm or facility aware of environmental issues and gives operational personnel a direct sense of participation in and ownership of environment-related issues.

- ISO 14001 offers a formal way to obtain and sustain the senior leadership's awareness of environmental issues and how they relate to the business as a whole.

- ISO 14001 supports a broad, quality-based approach to management that improves firm performance across the board. Put another way, ISO 14001 and this approach are synergistic. The quality-based approach includes environmental, health, and safety (EHS) integration and broader integration of all functions. In particular, ISO 14001 supports the cross-functional communication critical to such an approach and supports the continuous improvement seen as a key product of it.

- ISO 14001 promotes the recruiting and retention of labor by formally certifying the safety of the workplace.

So little time has passed since the introduction of ISO 14001 that it would be difficult to quantify the degree to which firms using it have realized their perceived benefits. Many firms believe the benefits they seek are so difficult to measure directly that they do not expect to have good quantitative measures of them, even in the future. They commit to ISO 14001, in whatever way they choose to use it, primarily on strategic grounds. That said, firms that have used ISO 14001 report benefits they had not expected to receive as often as they note benefits they expected but did not realize.

ISO 14001 AND THE DoD: AN ILLUSTRATION

A quick review of the benefits perceived by commercial firms verifies that some are more important to the DoD than others. For example, the DoD has little interest in adopting an EMS that protects market share relative to competitors. However, many of the benefits listed do appear to be relevant to the DoD.

A simple way to address this issue is to use ISO 14001 as a template to assess the effectiveness of a DoD EMS. A U.S. Air Force organization did exactly that with the initial version of ISO 14001.[6] Although the specification is slightly different today, no qualitative changes have occurred. A quick review of what the Air Force found illustrates how ISO 14001 relates to the DoD's current approach to environmental management and hence suggests the benefits that the DoD might realize from ISO 14001.

Table 4.2 summarizes the findings of a structured survey of Air Force environmental management professionals, who concluded that although the Air Force has not formalized its EMS in the exact way that ISO 14001 contemplates, the EMS can still achieve most of the goals of the ISO 14001 specification. In particular, the EMS reflects the strong preference of many government organizations for formal specification and documentation of their processes. ISO 14001 promotes this preference as long as the processes effectively link the organization's environmental concerns to its core strategic interests.

This comparison with the ISO 14001 specification suggests that the Air Force EMS can be improved in several specific ways:

- The Air Force EMS can move beyond its emphasis on compliance and give greater attention to a broader set of environmental aspects and their strategic implications for the performance of the Air Force as a whole. Such an emphasis can give continuous process improvement greater importance throughout the Air Force. ISO 14001 does not require such a change, but it provides an effective way to accomplish it if an organization is interested.

- The Air Force can give greater attention to the training, roles, and responsibilities of personnel who do not specialize in environmental activities.

- The Air Force can promote a greater commitment to a systemic approach to environmental management that traces environmental problems to their root causes, which often lie outside the immediate purview of the professional environmental staff.

- The Air Force can give more direct attention to monitoring and improving the EMS itself to reduce its cost of administration and improve its ability to link specific environmental aspects with broader strategic concerns.

---

[6]Lofgren et al., 1997.

Table 4.2

Assessment of a U.S. Air Force EMS in Relation to the ISO 14001 Specification

ISO 14001 Element	Strengths of USAF EMS	Areas Needing Improvement
Environmental policy		
General	Allows formulations of objectives and targets Commits to pollution prevention and compliance Clearly documented and communicated throughout environmental career field	Does not commit to continuous improvement Not clearly communicated outside environmental career field
Planning		
General	Formal methods to perform environmental impact statements Base natural resource and conservation plans help identify environmental impacts Weapon system design considers life-cycle environmental impacts	Driven by compliance, which does not motivate continuous improvement
Legal and other requirements	Higher directives and laws implemented through USAF environmental policy Bases identify site-specific environmental requirements Training and auditing cite specific environmental requirements	Could improve knowledge of requirements of local citizens, environmental groups
Objectives and targets	Set at the corporate level Consistent with USAF environmental policy Receive attention at every level	Driven by compliance Site-specific objectives and targets not universally established
Environmental management programs	Organization to realize objectives exists Objectives often include timelines	Responsibilities for objectives distributed among organizational levels not well defined

Table 4.2 (continued)

ISO 14001 Element	Strengths of USAF EMS	Areas Needing Improvement
Implementation and operation		
Structure and responsibility	Well-defined roles and responsibilities Responsibilities documented Full-time environmental management staffs Base environmental protection committees; unit environmental coordinators Adequate resource levels	Performance relative to EMS itself not reported to top leadership
Training, awareness, and competence	Environmental managers receive adequate training	Training focuses on environmental requirements, not environmental impacts Nonenvironmental personnel need more training
Communication	Automated management information system facilitates communication among environmental managers Public affairs can articulate issues to interested parties Public participates in restoration actions, environmental impact assessments	Procedures for external communication not fully developed Advisory boards could be applied beyond restoration
EMS documentation	USAF environmental policy and programs well documented	USAF EMS not documented
Document control	Strong document control	None
Operational control	Pharmacies manage hazardous maternals centrally, effectively Central agency reviews technical orders for environmental impacts Contracts specify environmental compliance requirements for suppliers Unit environmental coordinators monitor some unit operations	USAF should review specific operational procedures at work centers for environmental impacts

Table 4.2 (continued)

ISO 14001 Element	Strengths of USAF EMS	Areas Needing Improvement
Emergency preparedness and response	Base contingency plans include environmental responsibilities Bases practice emergency response Bases perform post-accident analyses to update procedures	None
Checking and corrective action		
Monitoring and measurement	Discharge permit standards measured continually Specific performance indicators and means to track them are in place Special system tracks status of all hazardous materials Performance indicators, internal inspections monitor compliance	Measures do not exist for all operations that can have environmental impact
Nonconformance and corrective and preventive action	Strong conformance mentality supports corrective action Someone is responsible for acting on audit findings and updating senior leadership Senior boards investigate some major incidents for root causes Bases share reports on corrective actions	Corrective actions do not always address systemic root causes
Records	Environmental records well managed Training records, position descriptions suitably maintained Correspondence with external agencies well recorded	None
EMS Audit	Compliance audits thorough and comprehensive Leaders get audit results as needed Procedures and responsibilities well documented	Limited to compliance Does not assess status of EMS itself Performance indicators track number, not nature, of findings
Management review		
General	Update performance indicators	EMS not audited and hence not explicitly reviewed

These findings are stated in broad terms here, but Table as 4.2 illustrates, ISO 14001's guidance is more pointed and targeted. The fact that ISO 14001 can offer specific insights is important in itself. It is also important that the opportunities for improvement so closely reflect insights from TQM. TQM emphasizes the importance of looking proactively beyond operating standards, integrating operations across functions, operating against a more systemic view of an organization's goals and potential, and self-consciously monitoring the organization's efforts to do these things. These actions are easy to list but very hard to accomplish. ISO 14001 provides a template that could help the Air Force—and by inference, all of the DoD—learn to become more proactive in environmental management.

The DoD is not alone. The observations in Table 4.2 are particularly important because they confirm important similarities between the DoD and large, proactive commercial firms. Even firms that have maintained and refined sophisticated EMSs since the 1970s typically find that the ISO 14001 specification helps them identify additional opportunities to improve. In particular, given the growing emphasis on a proactive approach in the 1990s, ISO 14001 helps even the best firms refine their EMSs in ways that promote a more proactive approach that better integrates environmental and strategic concerns.

IMPLEMENTATION

The most aggressive use of ISO 14001 employs a third-party auditor to register an organization to the standard. As noted above, this approach is still unusual in the United States. But reviewing the issues that American firms that have pursued this approach have encountered raises the implementation issues relevant to any approach to using ISO 14001. This section first considers the conditions present in an organization before it considers ISO 14001 that make success more likely. It then considers key challenges that organizations often confront in their implementation efforts. Finally, it outlines a nominal approach to implementation that anticipates such challenges and offers useful ways to deal with them.[7]

---

[7]An increasing number of implementation handbooks are appearing for ISO 14001. They provide much more detail on various aspects of the issues addressed here. See, for example, Cascio et al., 1996; Jackson, 1997a; and Woodside et al., 1998.

Preconditions That Promote Success

No firm approaches ISO 14001 in a vacuum. Preexisting circumstances in the firm help predict how easy it will be to adapt the firm's EMS to the ISO 14001 standard.

Regulatory or market preference for an ISO 14001-based EMS makes the largest, most immediate difference. To date, this has been relevant primarily outside the United States and probably by itself accounts for the greater willingness to register to ISO 14001 in Europe and Asia than in North America. This precondition is important in North American plants mainly to the extent that export markets affect corporate policy decisions. Different industrial firms view this kind of pressure quite differently. It is not particularly important to the DoD today.

Like the DoD, every large industrial organization must have a complex *EMS in place* simply to comply with all the regulations it faces. Firms typically use their EMS as the object of the first self-assessment. Typically, the more formal and proactive the EMS is, the more it looks like 14001, and hence the simpler registration to the standard is. The quality of EMSs often varies from division to division and from site to site within a firm. Hence, firms that use pilot sites to approach registration typically choose pilot sites with good preexisting EMSs.

Previous experience with a formal management system simplifies adaptation to the ISO 14001 standard. Possibilities include the ISO 9000 quality management system and its derivative, industry-specific standards. Previous pursuit of the Baldrige Award helps. Experience with an environmental system like EMAS, BS7750, or Responsible Care helps. Some self-selection occurs here; firms that like one standard will probably like another.

That said, ISO 9000 provides skills and procedures that a firm can use to build an EMS that reflects ISO 14001. Where possible, implementers involve their preexisting quality staff and processes. They often incorporate as much of their ISO 9000 systems as possible into the evolving EMS. Occasionally, firms use the development of a new EMS as an opportunity to incorporate lessons they have learned under ISO 9000. More often, they seem content to build on an imperfect version of ISO 9001 or 9002 simply to get the new EMS in place. They typically plan to refine it later.

Almost every firm that is able to do so ultimately plans to integrate its environmental and quality management systems so that they can draw on common modules and so that staff trained to work with one system can easily understand the other. Firms seek integration with comparable systems for managing safety and health as well. Such integration promotes compatible approaches to all management issues, making it easier to address the underlying goal of an overall quality approach—the integration of all process management with the core concerns of the firm.

The few firms that oppose such integration fear that it could compromise a still shaky ISO 9000-type certification. Very few think it would be easier to start without ISO experience; those that do so appear to want a fresh start that allows them to avoid their earlier mistakes, which have become imbedded in their existing quality management systems. Firms that use pilots to approach certification to ISO 14001 tend to favor pilot sites already registered to an ISO 9000-type standard.

Other Key Challenges to Implementation

Once decisionmakers in a firm decide to pursue ISO 14001, lack of early and persistent *senior leadership support* is easily the biggest problem they face. Many firms pursue ISO 14001 in a cumulative fashion, using the products of incremental implementation to build initial support for it and then building a growing case that the senior leadership must repeatedly endorse. Leaderships often accept a partial solution, allowing self-assessment and incremental steps to fill identified gaps but waiting for some external change to justify the final incremental cost of documenting implementation clearly enough to allow final registration.

The senior leadership typically wants to know how an EMS can advance the core interests of the firm. Because such core interests differ in public and private organizations, factors relevant to achieving and sustaining leadership support are also likely to differ. Common interests include lower costs, less downtime on the production line, fewer risks associated with continuing exposure to environmental risk, and clear explanations of how an EMS can be implemented without interrupting production. Private firms give a lot of attention to actual or anticipated customer demands. Formal documentation of the magnitude of such benefits can be a challenge, particularly early in the process; the EMS itself can potentially help with this if implementation proceeds incrementally, but in practice, clear documentation of improvements has been hard to achieve. The leadership must also understand why internal audits required by ISO 14001 will not subject a firm to additional regulatory liability that it would not face without ISO 14001.

The problem mentioned next most often is *availability of resources.* Smaller firms and sites, in particular, rely on employees who already have full-time responsibilities and opportunistically try to fit their work on a new EMS into gaps in their schedules where they are not already fully occupied. But any organization must recognize the added pressure on resources that implementation imposes, whether the resources are formally allocated to implementation or less formally squeezed from resources that are already fully employed. Constraints first hit the few people involved in designing the new EMS, then they hit those working on and responding to audits; ultimately, they affect everyone, as training consumes resources everywhere.

In the face of such constraints, related activities often provide the strongest competition for resources. For example, management of immediate compliance problems, which might themselves have first led a firm to think about a new EMS, will almost always have higher priority than an effort that cannot relieve these problems until some future date. Implementation of other new management systems can also take precedence. For example, a firm may favor implementation of an ISO 9000 standard because an EMS is easier to implement once a quality management system is in place, and the business case is generally stronger for ISO 9000 than for ISO 14001. A firm may favor a new management system for health and safety if employees face clear hazards. Senior leadership in general and headquarters support of a pilot site can relieve these constraints.

Support from staff beyond the environmental team is needed to form cross-functional process teams and then to adjust the production processes to reflect changes required to accommodate a new EMS.

This support is also critical to implementing specific instances of pollution prevention and waste minimization identified during the self-assessment that leads to the EMS. Such projects can help build the business case for the EMS. Accumulating empirical evidence of the value of commitment, pollution prevention, and waste minimization, coupled with motivational and operational training, helps sustain the interest and support of the nonenvironmental staff.

Managers and employees must understand how to think about environmental issues in terms of the firm's core concerns in a way that encourages continuing team-based management efforts that support and become part of the EMS. This typically takes time, experience with specific decisions or discussions between managers and employees, formal training, and repeated reinforcement of the corporate goals and incentives that support the application of the firm's core concerns to environmental issues.

Discretion in the design of an ISO 14001-based EMS leads to great ambiguity about how to implement such a system. Firms typically use teams and consensus-based systems to resolve ambiguity. Still, ISO 14001 allows a fair amount of latitude. In particular, firms must decide how much documentation is cost-effective within the constraints imposed by ISO 14001. Those that are unfamiliar with ISO-type standards can easily seek too much documentation, creating excess *ex ante* costs during implementation and continuing excess *ex post* costs associated with audits required to feed the procedures documented.

Firms implementing ISO 14001 voice a number of similar concerns about information they want but cannot find. Examples include the following:

- What are the best management practices to incorporate into an ISO 14001-based system? How can a firm find them?

- In the absence of a global EMS, how can a firm effectively isolate one site for the purposes of implementing an ISO 14001-based EMS? In particular, if products or processes are designed at site A and produced or used in manufacturing at site B, what responsibility does site B have for the design of these products or processes? If headquarters certifies a single recycler, reclaimer, or waste disposer for all sites, how do individual sites reflect this in their EMSs?

• Given the ambiguity associated with ISO 14001, how much is enough? When has a firm done enough that registration to the standard not only satisfies the auditor but also promotes the firm's core interests?

Consulting and auditing firms stand ready to provide advice on each of these questions. As a practical matter, the empirical experience available to address these questions remains limited in the U.S. setting, even though it is steadily accumulating over time. Frequently, expert firms draw on their past experience with ISO 9000-type systems. The quality of answers to these questions will inevitably improve as experience with ISO 14001 accumulates. Put another way, implementation should become easier over time as the leaders generate information about implementation that the organizations that follow can then exploit.

Firms differ on which part of ISO 14001 is the hardest to implement. Getting the basic environmental policy statement right is typically hard, and most firms also find the identification of significant environmental aspects hard. Some formally identify and consider stakeholders as they identify significant aspects. The aspects identified drive everything else and in particular determine the level of detail at which ISO 14001 will be defined, documented, and implemented. Once aspects are identified, defining goals, objectives, and targets is also challenging; this task often reopens issues thought to have been resolved earlier. These in turn raise specific issues about measurement, monitoring, and the availability of adequate data which involve generic challenges that previous experience with ISO 9000 can help address. As noted above, the specific challenges any organization faces depend on the nature of the EMS it starts with.

Almost invariably, documentation of existing processes is inadequate, even if the processes themselves are compatible with the ISO 14001 standard. Elements of ISO 14001 that appear to be easier to manage for firms with an ISO 9000 background include training and awareness, auditing, nonconformance, corrective and preventive action, records, document control, and management review. Firms can often transfer the procedures they developed for an ISO 9000 standard to ISO 14001 or even integrate them entirely. That said, maintaining effective training for auditors and controlling to keep auditors and documents up to date are difficult in either an ISO 9000 or an ISO 14001 setting.

A Nominal Implementation Process

Broadly speaking, ISO 14001 raises the same implementation issues raised by any large organizational change. Implementing ISO 14001 typically confirms the following imperatives:

- Build the case for change and sell it to the senior corporate and/or site leadership.

- Sustain appropriate support by the senior leadership, adequate resources, and effective cross-functional process teams.

- Execute an initial self-assessment.

- Execute an initial gap analysis that compares the current EMS with the ISO 14001-based target.

- Develop an overall action plan, with plans for each element of the EMS that identify appropriate roles and responsibilities, resources, and milestones. Execute the plan.

- Execute interim informal audits and adjustments.

- Identify and initiate cost-effective designs for the environmental, pollution-prevention, and waste-minimization actions relevant to but separate from the EMS itself. Use these to build incremental support for the EMS.

- Execute registration audits and final registration.

This kind of list could apply to many changes. Efforts to adopt an ISO 14001-based EMS typically focus on the key elements shown in Table 4.1. Every firm addressing ISO 14001 talks about these in slightly different terms. A firm's emphasis varies in proportion to its capabilities at the beginning of the process and the goals it associates with using ISO 14001. For example, Pitney Bowes focused on developing[8]

---

[8]Diamond, 1996.

- A detailed register of all environmental aspects.

- A process to rank environmental aspects.

- A register of relevant laws and regulations.

- Environmental targets and objectives based on the rankings above and company environmental policy.

- An EMS manual of procedures, including auditing, management review, and emergency response.

- Internal training for audits, with audits in progress.

- Employee awareness training.

- External communication plans.

Firms often seek external help to determine which elements to focus on. Some firms use the registrar for their ISO 9001 or 9002 as a consultant while developing ISO 14001, particularly if they seek integration. Given the degree of ambiguity inherent in ISO 14001, working with the final registrar from the beginning would seem to offer the most promising approach. But some firms prefer to use a third-party consulting firm to help develop their ISO 14001 system, and they then turn to a formal 14001 registrar for the necessary audits and final registration to the standard.

A firm must choose early on whether to pilot one or more sites, with plans to move beyond pilots or to implement ISO 14001 companywide. Going companywide without pilots is rare. Many firms choose a single-site pilot and focus there, deferring decisions about diffusion until later. Some even start with limited parts of the EMS at a site and build the system up over time. For example, one firm that RAND studied started with the environmental aspects easiest to measure and worked toward others later. This is most likely when the firm needs its relevant resources elsewhere and the EMS can develop only as free time becomes available. In a large organization, a limited pilot program can place too much burden on one site; a companywide approach can potentially spread the burden and let the firm draw on all its relevant assets during the initial implementation.

Firms use a broad range of approaches to the initial self-assessment that identifies gaps and initiates implementation. The only constant is the use of a team. That team can consist of from two to ten people at a site. It always includes at least one environmental manager, but it can also include the site manager, various functional people from the site (including production workers), and one or more people from headquarters. Those with previous ISO experience are typically ISO 9000 people, auditors or others. Contractors who specialize in EMS are also often included, but they are rarely the only ones involved in an initial assessment.

Implementation is not linear; everyone seems to benefit from several iterations of various kinds, in various elements. One firm reports that demonstrating hands-on success with a new process can help create momentum to support the tedious task of actually documenting the process; such a demonstration also obviously provides a test of the process, which could require changes before documentation. In general, repeated audits and reviews help to refine a consensus approach that yields the right level of detail as rounds of tests and documentation proceed. Many functional backgrounds and expert help—some of it external—are needed.

Training is critical to success. Some believe it is the most costly element of implementation, and its cost is often not captured because it resides primarily in the value of the time that employees spend in the training effort. Key elements of training include general motivation to be environmentally responsible (relevant to everyone); training on key elements of EMS, especially auditing (key personnel); and training on specific implications of the EMS for operations (everyone). Training continues to drive home the message, to explain changes in relevant laws and regulations and the EMS, and to train new personnel. Effective documentation and document control can limit the long-term cost of training by generating sources that employees can turn to on their own initiative for the information they need.

Stopping short of registration to the standard, with all systems ready to go to final registration as needed, is very common. This suggests that firms really do believe they can get ISO 14001 right without an external audit. This attitude must depend to some extent on the way implementation has proceeded.

In general, implementation looks easier in large organizations that can provide the necessary resources and bring specialized assets to bear where needed. Certainly, large firms appear to have played a greater role in the general development and application of TQM and TQEM than small firms have. But some organizations argue that implementation is easier in small firms because they are more agile and cross-functional communication is simpler. In all likelihood, large and small organizations benefit from different elements of ISO systems. Large firms get a structure that promotes the kind of internal communication that smaller firms, which are not as functionally splintered, take for granted. This relieves some scale diseconomies associated with large firms. Small firms get a cost-effective suite of management tools that they can use to keep up with changes in laws, regulations, stakeholder expectations, and technological opportunities. This suite reduces the scale economies associated with specialized management of these activities.

GLOBAL REGISTRATION TO ISO 14001 IN TWO LARGE AMERICAN FIRMS

The experience of two large American firms that successfully registered their major facilities worldwide to ISO 14001 provides valuable insights into these issues. The Ford Motor Company registered a variety of facilities with industrial activities whose environmental aspects are very much like those found in activities at DoD facilities that maintain aircraft and heavy vehicles. And IBM registered a variety of facilities with industrial activities whose environmental aspects are very much like those of activities at DoD facilities that maintain sophisticated electronic components. Like the DoD, both firms have facilities around the world. Both have long histories of proactive environmental management policies, but each chose a very different approach to achieving global registration.

Ford[9]

Ford is well known among American corporations for its commitment to environmental excellence. Its current environmental policy,

---

[9]This section is based on information from a variety of corporate sources, including the *1999 Environmental Report*, Ford Motor Company, Dearborn, MI, 1999. Also available, with related materials, at www.ford.com. Navigate from the home page to "Our Company," to "Better Ideas," to "Environment."

supported by the most senior leadership, dates from 1989. The CEO signed the Ford Environmental Policy, and the Environmental and Public Policy Committee of the Board of Directors oversees all environmental policies and practices. The Environmental Strategy Review Committee, co-chaired by the CEO and his chief of staff, designs an overall environmental leadership strategy and oversees its implementation in product development, in manufacturing, and elsewhere throughout the firm. The Vice President for Environmental and Safety Engineering implements this strategy day to day. This office uses the new Ford Environmental System, based directly on ISO 14001, to oversee environmental performance at 140 facilities worldwide. Ford's environmental policy drives continuous improvement relevant to every aspect of its business and ultimately contributes to the wellbeing of the firm and to its shareholder value.

Ford has had a formal EMS in place since the 1970s. Twenty years of experience yielded a sophisticated system; nevertheless, the new formal environmental management standards helped Ford fine-tune that system in ways that improved its environmental performance. At its own initiative, Ford began to reflect formal environmental management standards in its own EMSs at individual sites in Europe. For example, it implemented BS7750 at its Halewood, United Kingdom, Escort assembly plant and then the EMAS at its Saarlouis, Germany, Escort assembly facility. These plants both went on to register to ISO 14001.

The manager responsible for these actions, Graham Chatburn, moved to corporate world headquarters in Dearborn, Michigan, to become Ford's EMS manager. Within 15 months, in early 1996, he took on responsibility for registering 140 Ford facilities worldwide to ISO 14001. Ford sought a worldwide Ford Environmental System specifically designed to support registration to ISO 14001 and set an aggressive implementation schedule: six sites in Germany, Canada, the United Kingdom, and the United States by the end of 1996; 50 worldwide by the end of 1997; all 150 by the end of 1998.

Ford already had a commitment to ISO 9000 standards. It was also implementing the QS 9000 standards, derived from ISO 9000, with its suppliers, and it expected to realize benefits from ISO 14001 comparable to those it had already realized from ISO 9000. In particular:

- ISO 14001 would standardize Ford's EMS and related management processes worldwide. This would allow the Ford leadership to develop a single set of environmental requirements and goals and to focus on implementing them. The resulting commonality of effort would in turn support formalization of processes, which would be compatible with Ford's broader quality-based approach to management. It would also support Ford's efforts to make all of its employees aware of their responsibilities with regard to environmental performance.

- The resulting quality-based approach and employee awareness would support continuous improvement in environmental performance that should enhance operating efficiencies at individual plants and reduce waste and costs. In particular, it would identify hard metrics and use them to track improvements in performance over time. It would help Ford plants around the world share the best environmental practices, systematically reduce waste, and achieve greater operational efficiency.

- Competitive pressures to register to ISO 14001 are strong in Europe. Formal registration worldwide would give Ford a competitive advantage as it addressed these pressures.

- ISO 14001 would improve Ford's efforts to protect air, water, land, and other natural resources. Such improvements would enhance Ford's image with important stakeholder groups.

Ford also expected its past experience with ISO 9000 and QS 9000 to simplify its efforts to register to ISO 14001. In particular, it expected a great deal of overlap in documentation, procedures, and internal auditing processes. It also expected to use some of the cross-functional teams that worked on ISO 9000. With appropriate care, Ford would be able to use much of the registration process it had already developed.

Despite these expectations of built-in advantages, Ford charted an incremental approach to worldwide registration. In North America, it designated five plants that already had aggressively innovative track records as pilots:

1. Oakville, Ontario, assembly facility
2. Windsor, Ontario, aluminum facility

3. Lima, Ohio, engine facility

4. Van Dyke, Michigan, transmission facility

5. North Penn, Pennsylvania, electronics plant

Taken together, these plants spanned the types of manufacturing activities relevant to Ford's worldwide automobile operations.

Oakville was the first North American facility to register to ISO 9002; it has since registered to the more demanding ISO 9001. It then became the first automobile plant registered to ISO 14001 in North America. Oakville's implementation experience was invaluable in Ford's implementation of ISO 14001 at its Windsor, Ontario, and Markham, Ontario, electronics plants.

Oakville had already addressed its biggest environmental concern in 1992 by installing an automated paint shop that significantly reduced the use of cleaning solvents. Under ISO 14001, it expects to improve its sequencing of vehicle painting to further reduce pollutants, waste, and costs. It has also set specific targets to reduce its use of solvents, PCBs, energy, and nonreusable packaging.

Lima, Van Dyke, and North Penn were the first automobile plants in the United States to register to ISO 14001. The implementation of the ISO 14001-based Ford Environmental System at the Lima plant illustrates the kinds of challenges that many plants have experienced. The Lima implementation effort took about ten months. A cross-functional team steered the effort and helped the workforce understand that every one of the 2,100 employees in the 40-year-old plant had specific environmental responsibilities. This policy change involved significant culture change, which ISO 14001 facilitated by providing meaningful structure. For example, it helped the Lima employees do the following:

- Identify 30 significant environmental aspects, ranging from wastewater discharge and material handling to solid waste disposal. The employees identified objectives and targets to control, improve, or further study each of these aspects.

- Define new departmental work teams to implement and manage the new EMS.

- Systematically document their wealth of preexisting environmental knowledge relevant to procedures and work instructions. The documentation made this knowledge easily available to new personnel and other personnel considering process changes.

- Understand, through training and newly available supporting documentation, how the new EMS would work.

A series of audits provided an iterative process for implementation. An initial internal plant audit checked to see how well the new EMS was understood, implemented, and documented. Gaps were found and corrected. Then the plant management reviewed the status of the new EMS. Following a final preassessment review, the Vehicle Certification Agency (VCA) conducted an independent, third-party audit and registered the Lima plant to ISO 14001 in December 1996.

This total effort cost Lima about $220,000 in training costs and 5,600 employee-hours in meetings, training, and audits. Lima benefited from its previous registration to ISO 9001 and from the commitment of its employees to make the teams work. The Ford intranet also significantly simplified the development, maintenance, and distribution of accurate, up-to-date documentation. Lima later used the intranet to disseminate the lessons it had learned throughout the corporation.

Payoffs from the new system at Lima were immediate. Within a year of installing the new system, the Lima plant had

- Reduced water usage by 200,000 gallons per day.

- Eliminated production of boiler ash, the single largest component of the plant's solid waste stream.

- Increased the use of returnable packaging from 60 to 99 percent on its newest engine product (the AJ-30 V-8 engine for the Lincoln LS8).

Cross-functional teams of Ford employees have repeated efforts like those at Oakville and Lima at plants in 26 countries around the world with over 200,000 employees. This effort began in early 1996 and was completed by the end of 1998. Ford as a whole is already beginning to realize benefits from the kind of structured approach to process improvement that ISO 14001 supports. For example, Ford's 1998

Manufacturing Business Plan includes the following environmental-performance targets:

- Certify all manufacturing plants worldwide to ISO 14001 by the end of 1998 (achieved).

- Use 90 percent of returnable containers in facilities by 2001 (achieved in 1998).

- Reduce paint shop emissions by 60 grams per square meter by 2005.

- Phase out all PCB transformers by 2010.

- Reduce energy usage by 1 percent per year.

Ford currently uses its ISO 14001-based Ford Environmental System to track each of these goals, year by year. Each plant has its own additional, clearly stated objectives and targets.

IBM[10]

IBM is also well known among American corporations for its commitment to environmental excellence. It has had a formal corporate environmental policy since 1971, and it integrated this policy with its health and safety policies in 1990. IBM sees a direct connection between this commitment and its bottom line, and it tracks cost savings associated with environmental activities to maintain that connection. In recent years, these cost savings have tended to be about twice the amount that IBM spends each year on environmental activities.

IBM executives retain responsibility for the environmental performance of their organizations. The corporation's environmental programs govern all employees, not just those on environmental staffs. That said, the corporate environmental affairs staff sets IBM's worldwide environmental affairs strategy and monitors the implementation of this strategy. Environmental affairs personnel work at

---

[10]This section is based on information from a variety of corporate sources, including the *1998 Environmental Report,* International Business Machines Corporation, Armonk, New York, 1999. Also available, with related materials, at www.ibm.com/ibm/environment.

each manufacturing and R&D site as well as in major regional head-quarters. Environmental affairs focal points at various regional and operational headquarters support the integration of environmental policy with general business operations.

The corporate environmental program maintains a list of significant environmental aspects applicable to IBM products and of objectives and targets for these products on issues such as reuse and recyclability, upgradability, use of recycled materials, and improvement of energy efficiency. The Environmentally Conscious Products Strategy Owner for a product line then determines how these issues affect his or her product line. Teams for each product line address the centrally identified environmental issues and work with product designers to improve product performance relative to those issues. The product designers have final responsibility for each product's attributes but should be well informed about the environmental implications of their design choices.

IBM has used a sophisticated EMS for more than 25 years. Its Environmental Master Plan (EMP), the major planning and reporting document in the system, measures performance on chemical emissions and uses, effluents, waste management, recycling, and conservation at every manufacturing and R&D site in the company.[11] This reporting system helps IBM impose a common approach to environmental policy worldwide. The EMP provides data to IBM's senior leadership on a regular basis and helps business units analyze their own environmental performance and identify opportunities for improvement.

IBM took an immediate interest in ISO 14001 as the standard was being developed. IBM personnel actively participated in its refinement. Since ISO 14001 does not differ significantly from IBM's pre-existing EMS, IBM saw it as a natural complement to or extension of its ongoing efforts to maintain a position of environmental leadership in the corporate community. Registering to ISO 14001 would further demonstrate to relevant stakeholders the company's continuing commitment to environmental excellence. Early registration

---

[11]IBM uses a complementary system to collect environmental performance data on its administrative sites. These sites do not present the industrial challenges typical of the DoD's central logistics locations.

would also give IBM a competitive advantage if and when external stakeholders—regulators or customers—required such registration.

Despite strong similarities between ISO 14001 and the IBM EMS, an early gap analysis identified several places where ISO 14001 would improve effectiveness and efficiency and would further integrate environmental considerations around the company. The preexisting EMS managed environmental responsibility in the same manner at all IBM sites worldwide, so worldwide registration to the standard was expected to support the common framework and help drive common solutions to environmental management issues across sites. The resulting consistency and continual improvement would in turn yield increasingly efficient and effective management of environmental issues across the company.

IBM began to register individual sites, one at a time, to a draft version of ISO 14001 in 1996. Citing opportunities to achieve specific business advantages and to demonstrate a commitment to environmental excellence, five sites in the United Kingdom, Japan, and Singapore took the initiative to register quickly. (Other sites registered to EMAS as well.)

At the end of 1996, the first formal version of ISO 14001 allowed an organization to acquire a single worldwide registration for all of its sites. This approach allows an organization to start the registration process by selecting an appropriately large and representative sample of its sites and having them audited and registered. The organization can then add additional sites to the registration as they complete their audits and registrations. Diverse activities at different sites can be covered, as long as they all use the same EMS as a basis for the registration.

IBM saw that a single registration would play to the strengths of its preexisting EMS and would support further efforts to promote a common IBM approach to environmental management worldwide. A single registration would be more efficient and effective than continuing to register each site individually, and its size and sweep would dramatize IBM's continuing commitment to and unified global approach to environmental excellence.

IBM quickly decided to use this unified approach to register its corporate headquarters and 27 manufacturing and hardware R&D operations worldwide where the dominant environmental impacts occur (IBM has 33 such sites in all). These plants manufacture or develop microelectronics technology, data storage systems, personal computer systems, servers, and networking hardware. (IBM is the first firm to use this approach to worldwide registration on this scale, and it may add additional sites to the registration in the future.) IBM chose Bureau Veritas Quality International (BVQI), a registrar it had already worked with at many sites, to support its efforts to register all of these sites under a single registration. By December 1997, BVQI had registered eleven of the 28 sites, and all but two of the remaining sites were audited and registered by the end of 1998. Two more were covered in 1999. The sites registered before this unified effort began have migrated their registrations to the unified global registration.

IBM's commitment to environmental leadership includes doing business with environmentally responsible suppliers. It does not want to transfer responsibility for environmentally sensitive operations to companies that cannot manage them properly. Its contracts require that suppliers comply with all applicable laws and regulations in the work that they do with IBM. IBM gives special scrutiny to a select set of suppliers who present significant environmental risks or work primarily for IBM. Hazardous-waste and product-disposal vendors get special attention. IBM also shares technology with selected suppliers to improve their environmental performance.

IBM encourages its suppliers to align their EMSs to ISO 14001 and to pursue registration to the standard. Arguing from the perspective of an integrated supply chain, IBM expects ISO 14001 to improve the efficiency and effectiveness of its suppliers, and within an integrated supply chain, IBM could naturally expect some of those benefits to accrue to itself. That possibility aside, IBM reserves the right to require such registration in the future. If the image of an "environmentally responsible supplier" comes to include registration to ISO 14001, it will be natural for IBM to expect its suppliers to register. In contrast, Ford will in all likelihood focus on full implementation of QS 9000 among its suppliers before it gives serious attention to ISO 14001 in this context.

SUMMARY

Organizations can use ISO 14001 in many different ways to promote their environmental management goals. Within the United States, firms use it primarily as a template to assess the adequacy of their existing EMSs. Some do this as an end in itself, while others are preparing for the potential changes in market or regulatory circumstances that would make it attractive to register to the standard with an external auditor. However these firms approach ISO 14001, they perceive its relationship to TQM and the opportunity it presents to help them integrate their environmental management programs with their core business programs. Although the core interests of the DoD differ from those of the typical commercial firm, this aspect of ISO 14001 strongly suggests that it will be as useful to the DoD in many specific, identifiable ways as it has proven to be for many innovative commercial firms.

Implementation of ISO 14001 raises the same kinds of issues that arise in any large organizational change effort. Each organization faces implementation challenges that depend on, among other things, its environmental management goals; the quality of its existing EMS; its past experience with formal, quality-related management systems; the support of senior leadership; and the availability of resources. The standard itself allows considerable flexibility, giving organizations a great deal of leeway to implement it to pursue their own goals. That said, ISO 14001 provides enough structure to identify a core set of issues that any implementer must resolve. This in itself helps simplify the implementation and increases the likelihood that the implementation will lead to useful results.

Third-party auditors and consultants offer a wide range of services relevant to the implementation of ISO 14001. Because ISO 14001 is such a new standard, these firms are drawing on a limited body of empirical evidence on actual implementations, and this evidence will improve over time. No matter how good the information gets, however, ISO 14001 forces any implementer to reflect on its own goals and to craft its own processes to improve the integration of environmental management into its own broader business concerns.

Outside advisors can facilitate such an effort. But if ISO 14001 is to lead to real improvements in an organization, the implementer itself must give close attention to how it will use ISO 14001 on an ongoing basis.

The successful worldwide registrations to ISO 14001 of Ford and IBM help to illustrate what proactive organizations can do. With the strong and persistent support of their senior leaderships, and exploiting considerable staff experience with environmental and other formal management systems, both firms quickly transformed initial local registrations into global registrations for all of their major facilities. They did this to achieve a uniform environmental policy at all locations that would simplify corporate management of environmental policy, but each chose a very different way to move from local to global registration. Each implementation was tailored to particular circumstances, and each worked well in its own setting.

Chapter Five

CONCLUSIONS

Lessons from environmental management in commercial settings can be applied to DoD central logistics activities. In wholesale-level transportation and supply, fuel consumption and packaging are particularly important environmental issues, but many other concerns also arise. Depot-level maintenance of defense systems raises the same variety of environmental issues that exist in the manufacture of comparable integrated electronics and aircraft systems, engines, and heavy equipment. These issues range from direct emissions to hazardous-waste generation. No one environmental concern dominates in central logistics. The sources of environmental problems and, by implication, the keys to solving them exist throughout the processes that produce central logistics services. They are, for practical purposes, an integral part of these processes. Any effort to manage these concerns must recognize how intertwined environmental issues are with the design and operation of logistics production processes.

A broad consensus is emerging from the many commercial-sector experiments in proactive environmental management that are under way today. Many commercial firms with production processes analogous to those in central logistics are experimenting with proactive environmental management. Each year, additional major firms find ways to look beyond simply complying with current regulations. They cite many reasons for doing this. As regulations have become more pervasive, more and more cost-effective opportunities to move away from end-of-the-pipe controls to pollution prevention present themselves. Planning before regulations are imposed gives an organization a better opportunity to develop a cost-effective response to such regulations. As global entities, large firms find it easier to develop a single, organizationwide policy than to worry, at the corporate level, about specific compliance issues in each locale. As highly visible parts of the communities where their facilities reside, they can affirm their status as responsible citizens by doing more than the regulations require.

How successful have these experiments been? It is too early to use survival analysis or other formal techniques to determine how the current experiments affect the business performance of the firms conducting them. But our analysis clearly indicates that interest in such experiments continues to grow among large commercial firms. And their stated reasons for conducting them are clearly relevant to the DoD. Each firm finds a way to tailor its environmental management approach to fit its own corporate culture. To the extent that the DoD wants to emulate these firms, it too will need to fit its environmental programs to its own culture.

The environmental management policy that the DoD promotes today is broadly compatible with the consensus emerging from best commercial practice. The DoD's environmental policy highlights many of the points that proactive commercial firms emphasize: Develop and maintain good working relationships with key stakeholders; develop clear organizational goals and metrics to support them; improve information systems that can be used to monitor performance; motivate the workforce and empower it with effective training and analytic tools; support experimentation with innovative technologies; use affirmative acquisition policies to encourage suppliers to be more environmentally responsible; and so on. From this perspective, high-level environmental policymakers in the DoD already support much of the consensus forming in the commercial sector.

The DoD should integrate environmental management with its core mission concerns. Integration makes connections in both directions. Moving corporate values into the realm of environmental management emphasizes the importance of going beyond compliance only to the extent that such a change is compatible with the corporate strategy. Moving an environmental perspective into the core production processes of the firm creates new opportunities for these processes to pursue the long-term corporate strategy. Effective integration is achieved when environmental concerns are no more and no less important than the other concerns in an organization. This view is compatible with the observation that environmental concerns are integrally intertwined with the core concerns of designing and operating central logistics production processes.

Such integration is natural in central logistics activities. The DoD is already using an integrated supply chain perspective to link logistics functions to one another and to the core military missions of the department. Agile combat support and velocity management are two innovative examples of this ongoing effort. Integrated supply chain management provides precisely the perspective needed to bring environmental concerns closer to the core concerns of DoD logistics. We believe that it provides a natural way to ask, for example, where shortened cycle times are compatible with environmental and core military goals and where they are not. It provides a natural way to ask where pollution-prevention activities can reduce emissions and waste in ways that reduce systemwide operating costs inside the DoD. In sum, current DoD thinking about integrated logistics services provides a natural conduit for promoting the integration of environmental and core concerns.

But pursuing an integrated supply chain is not enough. Commercial practice suggests that environmental specialists must have high enough visibility for the managers of core logistics activities to take them seriously. Environmental specialists must advocate the environmental position, but they must do so in language that logisticians can appreciate. Environmental specialists must bring information about the regulatory consequences of logistics practices to the attention of those with responsibility for changing those practices. Ideally, environmental specialists and logisticians—managers, production workers, and analysts—should work face to face at multiple levels and develop good working relationships. Cross-functional teams can help this happen, but they are likely to succeed only if their members are trained to pursue consensus and have the authority to commit their individual functions to team-developed solutions. The DoD has a great deal to learn from commercial practice here.

Many commercial firms speak broadly about how they use environmental opportunities to promote strategic competitiveness. However, the most compelling specific examples of moving beyond compliance justify themselves in dollars and cents. They depend on an approach to accounting that captures all of the costs relevant to environmental concerns and brings them to bear on the decision at hand. Viewed in this way, these examples apply activity-based costing (ABC). Very few formal ABC accounting systems exist, however, so the DoD needs to learn how proactive commercial firms are using their existing accounting systems to gather the data they apply to environmental decisions.

Environmental policy is only one area in which the DoD needs to learn how to apply ABC. Developing the approach to costing needed to support proactive environmental management should be just one part of DoD's broader efforts to adopt ABC throughout its operations and investment decisionmaking. ABC is ultimately just a tool that can help the DoD integrate environmental concerns with its other concerns. The DoD should apply ABC to environmental issues, logistics issues, and other issues in an integrated, internally consistent way. It should pursue an incremental approach to ABC analysis, using early successes with simple versions of it to build department-wide support for a variety of challenging applications. Environmental issues offer good places to start this process.

A formal environmental management program can increase the likelihood that implementation of proactive environmental management will succeed. Even as a proactive organization integrates environmental management and management of its core mission, it relies on a formal environmental program to maintain focus. The program cannot directly do the things that make the program succeed; core production activities must do that. Rather, the program maintains visibility over environmental issues to ensure that core production activities take them seriously.

A "formal program" typically includes managers whose sole responsibility is environmental policy at the corporate, business-unit, and plant level. It includes a staff to support these managers and a career track to develop staff and managers over time. This career track provides the specialists who serve on cross-functional teams.

A formal program also typically includes a monitoring system that develops goals and objectives at each level in the organization, develops metrics to monitor performance against those objectives, and routinely reports information on those metrics to the leaders re-sponsible for core production activities at each level of the organiza-tion. Environmental specialists help the core production personnel develop these goals, objectives, and metrics and the incentives to support them, but the core personnel are ultimately responsible for them. The monitoring system provides a way for the leadership to see how well the core activities execute against their own plans and to reward them accordingly. The environmental program provides support and advice in this effort, but the environmental specialists do not make decisions about production activities, investments, or rewards.

Among other things, a formal program monitors the development and execution of proactive environmental initiatives. It monitors progress against the milestones developed for each initiative, and it helps structure the set of initiatives itself to promote learning over time. Successful firms typically try easy, lower-risk initiatives first, i.e., actions farther from core production activities. These firms build on success. They use early success to maintain corporate support for more work and to develop lessons that the organization can use to attack more difficult challenges. These firms "seed" initiatives at the beginning to promote such learning in each of their business units and key production areas.

In the context of the DoD, most of the work in such a program would occur in the military departments and agencies and, within them, probably at the major command level. OSD would reserve for itself the (very demanding) tasks of maintaining the support of the senior DoD leadership, helping that leadership develop DoD-wide envi-ronmental goals, and reporting to the leadership on a regular basis on progress toward those goals. The program should be institu-tionalized to recognize the likelihood that initiatives that began under one leadership team would very often reach completion under a successor.

Total Quality Management (TQM) provides useful formal templates that the DoD can use to verify its approach to implementing pro-active environmental management. Many proactive firms em-phasize the inherent compatibility of TQM and proactive environ-mental management, particularly pollution prevention. TQM and pollution prevention share the goal of eliminating waste—any activity that does not contribute value for the final customer or stakeholder. Such a bottom-up perspective is especially appropriate in central logistics activities, where waste exists throughout the supply chain and can be flushed out only by a systematic effort to review each process and its contribution to final customer value. In this setting, pollution prevention becomes just one point of focus for the TQM continuous improvement process at all levels. Seen in this way, TQM creates another opportunity to link an organization's environmental management processes with its core mission processes.

Several formal TQM-based templates provide the basis for monitoring an organization's efforts to create and maintain a formal environmental management program like that described above, from the top down. ISO 9000 and ISO 14000 offer compatible frameworks for creating a quality-based organization and making sure that it reflects environmental management concerns effectively. Used together, these standards promote the kind of integration discussed above. The Baldrige Award criteria and TQEM matrix developed by the Council of Great Lakes Industries provide mutually compatible frameworks for moving beyond the requirements of the ISO templates. Again, used together, they promote integration in an organization seeking best-of-class status, not just in environmental management, but across the board.

Many organizations use TQM methods for environmental management without adopting these formal templates. The tools exist to help organizations that cannot implement a quality-based management approach without them. Many commercial firms have recognized this, using ISO 9000 to get started and moving beyond the ISO requirements as they gain confidence and experience. The DoD currently uses TQM in many individual locations but has rarely adopted even the most basic of the formal templates, ISO 9000. Is the DoD ready to take a more formal approach to TQM, in environmental management and elsewhere?

Effective, proactive environmental management in central logistics activities could help lead the DoD toward broad acceptance of TQM. As noted above, a common theme in commercial experience is the need to develop an approach to environmental management that is compatible with the core values of the organization. Firms that have adopted TQM naturally saw it as a tool they could use to pursue proactive environmental management. In fact, some saw the application of TQM to environmental issues precisely as a natural way to justify environmental efforts in their corporate structures and hence a natural way to integrate environmental concerns with the broader concerns of the firm. TQM became the natural vehicle to use to promote integration.

The DoD does not maintain a quality-based infrastructure. The functional organizations in the services that maintain TQM skills can point to many successes, but they have not penetrated the core activities of the services. As the DoD reviews its infrastructure to squeeze out the dollars it needs to recapitalize the force in the new century, it will have to achieve dramatic improvements in performance. TQM and related reengineering perspectives are the only concepts that have yielded such improvements in commercial firms. If the DoD is to improve its infrastructure as it believes it must, TQM will have to move toward center stage soon.

DoD environmental policymakers can take advantage of that move by linking environmental management to TQM in a way that puts them in common cause with innovators elsewhere in the department. TQM offers the potential for creating the centerpiece for a coalition of innovators who work together to change the DoD's approach to managing the infrastructure. Existing proactive environmental policy in the DoD has already created an opening. Environmental policy might actually serve as an ideal starting place to insert wedges to work toward such an infrastructure. Because environmental concerns are more distant from the core concerns of the DoD than most infrastructure issues, they offer a low-risk place to test TQM concepts that could be applied elsewhere in the department if they prove to be successful.

Central logistics activities, themselves more distant from the DoD's core concerns than many other activities, could offer an ideal place to test these ideas. Planting the seeds of TQM in the environmental management of central logistics could help provide the basis for the much broader success the DoD must experience to draw down its infrastructure without sacrificing military capability. Such seeds could promote integration, not by bringing environmental management into the fold of an existing TQM corporate culture, but by using environmental management to help jump-start the change that will move all of the DoD into a TQM culture.

Bailey, Paul E., and Peter A. Soyka, "Making Sense of Environmental Accounting," *Total Quality Environmental Management,* Spring 1996a, pp. 1–15.

_____, "Environmental Accounting—Making It Work for Your Company," *Total Quality Environmental Management,* Summer 1996b, pp. 13–30.

Begley, Ronald, "ISO 14000: A Step Toward Industry Self-Regulation, *Environmental Science and Technology,* Vol. 30, No. 7, 1996, pp. 298–302.

Berube, Michael, et al., "From Pollution Control to Zero Discharge: How the Robbins Company Overcame the Obstacles," *Pollution Prevention Review,* Spring 1992, pp. 189–207.

Blumenfeld, Karen, and Anthony Montrone, "Environmental Strategy—Stepping Up to Business Demands," *Prism,* Fourth Quarter 1995, pp. 79–90.

Breton, Hal, et al., "Case Study: AT&T Is Setting and Achieving Stretch Goals," *Pollution Prevention Review,* Autumn 1991, pp. 389–403.

Brown, Howard, and Jim Dray, "Where the Rubber Meets the Road: Measuring the Success of Environmental Programs," *Total Quality Environmental Management,* Spring 1996, pp. 71–80.

Business Roundtable, *Facility Level Pollution Prevention Benchmarking Study,* Washington, DC, 1993.

Butner, Scott, "ISO 14000—Policy and Regulatory Implications for State Agencies," National Pollution Prevention Roundtable, Spring National Meeting, Washington, DC, April 1996a, pp. 1–8.

_____, "Web Sites Show Greening of Corporations," *Seattle Daily Journal of Commerce*, August 22, 1996b.

Cairncross, Frances, *Costing the Earth*, Boston: Harvard Business School Press, 1991.

_____, "How Europe's Companies Reposition to Recycle," *Harvard Business Review*, March–April 1992, pp. 38–47.

Camm, Frank, *The Development of the F100-PW-220 and F110-GE-100 Engines: A Case Study of Risk Assessment and Risk Management*, N-3618-AF, Santa Monica, CA: RAND, 1993.

_____, *Expanding Private Production of Defense Services*, MR-734-CRMAF, Santa Monica, CA: RAND, 1996.

Cascio, Joseph, and Gregory J. Hale, "ISO 14000: A Status Report," *Quality Digest*, February 1998.

Cascio, Joseph, Gayle Woodside, and Philip Mitchell, *ISO 14000 Guide: The New International Environmental Standards*, New York: McGraw-Hill, 1996.

Clarke, Richard A., et al., "The Challenge of Going Green," *Harvard Business Review*, July–August 1994, pp. 37–50.

Commission on Roles and Missions of the Armed Forces, *Directions for Defense*, Final Report, Washington, DC: U.S. Government Printing Office, 1995.

Defense Science Board, *Report of the Defense Science Board Task Force on Environmental Security*, Washington, DC: Office of the Under Secretary of Defense (Acquisition and Technology), 1995.

_____, *Report of the Defense Science Board Task Force on Outsourcing and Privatization*, Washington, DC: Office of the Under Secretary of Defense (Acquisition and Technology), 1996a.

_____, *Achieving an Innovative Support Structure for Twenty-First Century Military Superiority: Higher Performance at Lower Costs,*

Washington, DC: Office of the Under Secretary of Defense (Acquisition and Technology), 1996b.

DeLange, Cecil, "Interdepartmental Teamwork Results in Innovative Solution to Environmental Problem," *Industrial Engineering,* August 1994, pp. 54–55.

Diamond, Craig P., *Environmental Management System Demonstration Project: Final Report,* Ann Arbor, MI: NSF, Inc., December 1996.

Dobyns, Lloyd, and Clare Crawford-Mason, *Quality or Else: The Revolution in World Business,* Boston: Houghton Mifflin, 1991.

Drezner, J., and F. Camm, *Using Process Design to Improve DoD's Environmental Security Program: Remediation Program Management,* MR-1024-OSD, Santa Monica, CA: RAND, 1999.

Duke, L. Donald, "Pollution Prevention and Hazardous Waste Management in Two Industrial Metal Finishing Facilities," *Hazardous Waste and Hazardous Materials,* Vol. 11, No. 3, 1994, pp. 435–457.

Dumond, John, et al., *Velocity Management: An Approach for Improving the Responsiveness and Efficiency of Army Logistics Processes,* DB-126-A, Santa Monica, CA: RAND, 1994.

Ehrenfeld, John R., and Jennifer Howard, "Setting Environmental Goals: The View from Industry—A Review of Practices from the 60s to the Present," Draft, MIT Program on Technology, Business, and Environment, Cambridge, MA: Massachusetts Institute of Technology, 1995.

European Recovery and Recycling Association, *ERRA Principles of Action: A Comprehensive Household Recovery and Recycling System,* Brussels: European Recovery and Recycling Association, 1991.

Ferrone, Bob, "Environmental Life-Cycle Management Emerges," *Total Quality Environmental Management,* Spring 1996, pp. 107–112.

Freeman, H., et al., "Industrial Pollution Prevention: A Critical Overview," *Journal of the Air and Waste Management Association,* Vol. 42, 1992, pp. 618–639.

Girardini, Kenneth, et al., *Improving DoD Logistics,* DB-148-CRMAF, Santa Monica, CA: RAND, 1995.

Global Environmental Management Initiative, *Finding Cost-Effective Pollution Prevention Initiatives: Incorporating Environmental Costs into Business Decision Making—A Primer,* Washington, DC, 1994.

_____, *Global Environmental Management Initiative, The Primer,* Washington, DC, 1992.

"The Green Machine: Environmental Regulations and Responsibilities Becoming More Important," *Quality Progress,* March 1995, pp. 17–18.

Greeno, J. Ladd, et al., "Rethinking the Environment for Business Advantage," *Prism,* First Quarter 1996, pp. 5–15.

Grogan, Peggy, "Preventive Maintenance: A Valuable P2 Tool," *Pollution Prevention Review,* Spring 1996, pp. 53–62.

Haines, Robert W., "Environmental Performance Indicators: Balancing Compliance with Business Economics," *Total Quality Environmental Management,* Summer 1993, pp. 367–372.

Hoffman, Andrew, "Teaching Old Dogs New Tricks: Creating Incentives for Industry to Adopt Pollution Prevention," *Pollution Prevention Review,* Winter 1992/93, pp. 1–11.

Hunter, J. S., and D. M. Benforado, "Life Cycle Approach to Effective Waste Minimization," *Proceedings, 80th Annual Meeting,* Air and Waste Management Association, 1987.

Jackson, Suzan L., "Certification of Environmental Management Systems—for ISO 9000 and Competitive Advantage," in John T. Willig (ed.), *Auditing for Environmental Quality Leadership,* New York: John Wiley and Sons, 1995, pp. 61–69.

_____, *The ISO Implementation Guide: Creating an Integrated Management System,* New York: John Wiley and Sons, 1997a.

_____, "ISO 14000: Things You Should Know," *Automotive Manufacturing and Production*, October 1997b.

Johnson, Perry, *ISO 9000: Meeting the New International Standards*, New York: McGraw-Hill, 1993.

Johnson, Thomas H., and Robert S. Kaplan, *Relevance Lost: The Rise and Fall of Management Accounting*, Boston: Harvard Business School Press, 1987.

Junkins, Jerry R., "Insights of a Baldrige Award Winner," *Quality Progress*, March 1994, pp. 57–58.

Juran, J. M., "The Upcoming Century of Quality," *Quality Progress*, August 1994, pp. 64–71.

Kainz, Robert J., Monica H. Prokopyshen, and Susan A. Yester, "Life Cycle Management at Chrysler," *Pollution Prevention Review*, Spring 1996, pp. 71–83.

Kaplan, Robert S. (ed.), *Measures for Manufacturing Excellence*, Boston: Harvard Business School Press, 1990.

Kaplan, Robert S., and David P. Norton, *The Balance Scorecard: Translating Strategy into Action*, Boston: Harvard Business School Press, 1996.

Kirchenstein, John J., and Rodger A. Jump, "The European Eco-label and Audit Scheme: New Environmental Standards for Competing Abroad," in John T. Willig (ed.), *Auditing for Environmental Quality Leadership*, New York: John Wiley and Sons, 1995, pp. 70–78.

Kirschner, Elisabeth, "Environmental Managers Chip at 'Green Wall,'" *Chemical and Engineering News*, February 26, 1996, pp. 19–20.

Klafter, Brenda, "Pollution Prevention Benchmarking: AT&T and Intel Work Together with the Best," *Total Quality Environmental Management*, Autumn 1992, pp. 27–34.

Klafter, Brenda A., et al., "Environmental Benchmarking: AT&T and Intel's Project to Determine the Best-in-Class Corporate Pollution Prevention Programs," *Corporate Quality/Environmental Man-*

agement III: Leadership—Vision to Reality, Conference Proceedings, Arlington, VA, March 24–25, 1993, pp. 37–43.

Krut, Riva, and Carol Drummond, *Global Environmental Management: Candid Views of Fortune 500 Companies,* Washington, DC: U.S.-Asia Environmental Partnership, 1997.

Lagoe, David J., "Case Study: Alcan Rolled Products Company," *Pollution Prevention Review,* Winter 1995/96, pp. 51–57.

Lent, Tony, and Richard P. Wells, "Corporate Environmental Management Study Shows Shift from Compliance to Strategy," *Total Quality Environmental Management,* Summer 1992, pp. 372–394.

Leonard-Barton, Dorothy, *Wellsprings of Knowledge,* Boston: Harvard Business School Press, 1996.

Levine, Arnold, and Jeff Luck, *The New Management Paradigm: A Review of Principles and Practices,* MR-458-AF, Santa Monica, CA: RAND, 1994.

Lofgren, Steven T., et al., "An Examination of the U.S. Air Force Environmental Management System in Relation to the International Organization for Standardization (ISO) 14001 Specification," Presentation 97-TP36B.07, Air and Waste Management Association, 90th Annual Meeting and Exhibition, Toronto, Canada, June 8–13, 1997.

Loren, James, "Weyerhaeuser Implements New Chemical Management Program," *Total Quality Environmental Management,* Spring 1996, pp. 101–106.

Malachowski, Mark, "Case Study: Hewlett-Packard's Sunnyvale Facility Is a Leader in Waste Minimization," *Pollution Prevention Review,* Autumn 1991, pp. 405–410.

Marchetti, John A., et al., "Overcoming the Barriers to Pollution Prevention," *Pollution Prevention Review,* Winter 1995/96, pp. 41–50.

Maxwell, James, et al., "Case Study: Preventing Waste Beyond Company Walls: P&G's Response to the Need for Environmental Quality," *Pollution Prevention Review,* Summer 1993, pp. 317–333.

McCully, Annette Dennis, "Unlikely Partners: How QS-9000 and ISO 14000 Work Together," *Quality Digest*, 1998.

Megna, Alice T., and Tim Savoy, "Case Study: Amoco Production Company—Training for Pollution Prevention, *Pollution Prevention Review*, Autumn 1991, pp. 411–417.

Minner, Jill F., "In pursuit of ISO14000," *Quality*, January 1997.

Nagel, George, "Business Environmental Cost Accounting Survey," in Global Environmental Management Initiative, *Environmental Management in a Global Economy*, Washington, DC, 1994, pp. 243–248.

Ochsner, Michele, Caron Chess, and Michael Greenberg, "Case Study: DuPont's Edgemoore Facility," *Pollution Prevention Review*, Winter 1995/96, pp. 65–74.

O'Dea, Katherine, and Gregg Freeman, "Environmental Logistics Engineering: A New Approach to Industrial Ecology," *Total Quality Environmental Management*, Summer 1995, pp. 73–85.

Owen, Jean V., "The Greening of the Manufacturing Engineer," *Manufacturing Engineering*, March 1995a, p. 4.

_____, "Environmental Compliance: Managing the Mandates," *Manufacturing Engineering*, March 1995b, pp. 59–66.

Palmer, Adele R., et al., *Economic Implications of Regulating Chlorofluorocarbon Emissions from Nonaerosol Applications*, R-2524-EPA, Santa Monica, CA: RAND, 1980.

Paveley, Hilary, "Logistics and the Environment," *Logistics Today*, September/October 1992, pp. 24–26.

Penman, Ivy, and James R. Stock, "Environmental Issues in Logistics," *The Logistics Handbook*, New York: Free Press, 1994, pp. 840–857.

Perry Johnson, Inc., *QS-9000: An Executive Overview*, Southfield, MI, 1995, cited in "QS-9000: Quality," *Automotive Engineering*, June 1995, pp. 61–65.

Pfaff, Alexander S. P., and Chris William Sanchirico, "Environmental Self-Auditing: Setting the Proper Incentives for Discovery and Correction of Environmental Harm," Draft, New York: Columbia University, 1999.

Piasecki, Bruce W., *Corporate Environmental Strategy: The Avalanche of Change Since Bhopal*, New York: John Wiley and Sons, 1995.

Pojasek, R., "Waste Reduction Audits," *Environmental Risk Management—A Desk Reference*, Alexandria, VA: RTM Communications, 1991.

Porter, Michael E., and Claas van der Linde, "Green and Competitive: Ending the Stalemate," *Harvard Business Review*, September–October 1995, pp. 120–134.

Quinn, Barbara, "Creating a New Generation of Environmental Management," *Pollution Engineering*, June 1997.

Ramey, Tim, *Lean Logistics: High-Velocity Logistics Infrastructure and the C-5 Galaxy*, MR-581-AF, Santa Monica, CA: RAND, 1999.

Rank Xerox, *Environmental Performance Report*, London, 1995.

Resetar, S., F. Camm, and J. Drezner, *Environmental Management in Design: Lessons from Volvo and Hewlett-Packard for the Department of Defense*, MR-1009-OSD, Santa Monica, CA: RAND, 1998.

Richards, Deanna J., and Robert A. Frosch (eds.), *Corporate Environmental Practices: Climbing the Learning Curve*, Washington, DC: National Academy Press, 1994.

Ritchie, Ingrid, and William Hayes, *A Guide to the Implementation of the ISO 14000 Series on Environmental Management*, New York: Prentice Hall, 1997.

Rose, John, and Graham Sharman, "The Redesign of Logistics," *McKinsey Quarterly*, Winter 1989, pp. 29–43.

Rubenson, David, Jerry Aroesty, and Charles Thompson, *Two Shades of Green: Environmental Protection and Combat Training*, R-4220-A, Santa Monica, CA: RAND, 1992.

Rubenson, David, et al., *Marching to Different Drummers: Evolution of the Army's Environmental Program*, MR-453-A, Santa Monica, CA: RAND, 1994.

Schmidheiny, Stephan, *Changing Course: A Global Business Perspective on Development and the Environment*, Cambridge, MA: MIT Press, 1992.

Sheldon, Christopher (ed.), *ISO 14001 and Beyond: Environmental Management Systems in the Real World*, Sheffield, UK: Greenleaf Publishing, 1997.

Spriggs, H. Dennis, "Integration: The Key to Pollution Prevention," *Waste Management*, Vol. 14, Nos. 3–4, 1994, pp. 215–229.

Stapleton, Philip J., Anita M. Cooney, and William M. Hix, Jr., *Environmental Management Systems: An Implementation Guide for Small and Medium-Sized Organizations*, Ann Arbor, MI: NSF International, 1996.

Starik, Mark, Alfred A. Marcus, and Anne Y. Ilinitch, "Special Research Forum: The Management of Organizations in the Natural Environment," *Academy of Management Journal*, Vol. 43, No. 4, August 2000, pp. 539–736.

Stillwell, E. Joseph, et al., *Packaging for the Environment: A Partnership for Progress*, New York: AMACOM, 1991.

Stock, James R., *Reverse Logistics*, Oak Brook, IL: Council of Logistics Management, 1992.

Stratton, Brad, "Goodbye, ISO 9000; Welcome Back, Baldrige Award," *Quality Progress*, August 1994, p. 5.

Thompson, Barbara C., and Albery C. Rauck, "Applying TQEM Practices to Pollution Prevention at AT&T's Columbus Works Plant," *Total Quality Environmental Management*, Summer 1993, pp. 373–381.

Turbide, David A., "Japan's New Advantage: Total Productive Maintenance," *Quality Progress*, March 1995, pp. 121–123.

Ummenhofer, Matthais, "Managing Material Flows in a Complex Business Environment," *Logistics Spectrum*, Spring 1995, pp. 24–29.

U.S. Bureau of the Census, *Statistical Abstract of the United States, 1996*, 116th ed., Washington, DC, 1996.

U.S. Environmental Protection Agency, *Facility Pollution Prevention Guide*, EPA/600-R-92-088, Washington, DC: Office of Research and Development, 1992.

_____, *Federal Facility Pollution Prevention Planning Guide*, EPA/300-R-94-013, Washington, DC: Office of Enforcement and Compliance Assurance, 1994.

_____, *Profile for the Electronics and Computer Industry*, EPA/310-R-95-002, Washington, DC, 1995a.

_____, *Profile for the Motor Vehicle Assembly Industry*, EPA/310-R-95-009, Washington, DC, 1995b.

_____, *Prototype Study of Industry Motivation for Pollution Prevention*, EPA 100-R-96-001, Washington, DC: Office of the Administrator, 1996a

_____, *Prototype Study of Industry Motivation for Pollution Prevention: Interview Results*, EPA 100-R-96-001a, Washington, DC: Office of the Administrator, 1996b.

Van Laarhoven, P., and Graham Sharman, "Logistics Alliances: The European Experience," *McKinsey Quarterly*, No. 1, 1994.

Walley, Noah, and Bradley Whitehead, "It's Not Easy Being Green," *Harvard Business Review*, May–June 1994, pp. 46–52.

Wever, Grace H., "Engaging Your Networks," *Total Quality Environmental Management*, Summer 1996a, pp. 31–39.

_____, *Strategic Environment Management: Using TQEM and ISO 14000 for Competitive Advantage*, New York: John Wiley and Sons, 1996b.

Wever, Grace H., and George F. Vorhauer, "Kodak's Framework and Assessment Tool for Implementing TQEM," *Total Quality Environmental Management*, Autumn 1993, pp. 19–30.

Williams, Marcia E., "Why—and How to—Benchmark for Environmental Excellence," *Total Quality Environmental Management*, Winter 1992/93, pp. 177–185.

Willig, John T. (ed.), *Environmental TQM*, 2nd ed., New York: McGraw-Hill, 1994.

_____, *Auditing for Environmental Quality Leadership*, New York: John Wiley and Sons, 1995.

Wilson, Robert C., "ISO 14004 Can Alleviate EMS Growing Pains," *Pollution Engineering*, January 1998.

Wolf, Kathleen, and Frank Camm, *Policies for Chlorinated Solvent Waste—An Exploratory Application of a Model of Chemical Life Cycles and Interactions*, R-3506-JMO/RC, Santa Monica, CA: RAND, 1987.

Womack, James P., and Daniel T. Jones, *Lean Thinking*, New York: Simon and Schuster, 1996.

Woods, Sandra K., "Making Pollution Prevention Part of the Coors Culture," *Total Quality Environmental Management*, Autumn 1993, pp. 31–38.

Woodside, Gayle, Patrick Aurrichio, and Jeanne Yturri, *ISO Implementation Manual*, New York: McGraw-Hill, 1998.